Basic Computer Logic

Lexington Books Series in Computer Science

Kenneth J. Thurber, General Editor

Basic Computer Logic
John R. Scott
Computing in the Humanities
Peter C. Patton and *Renee A. Holoien*
Data Base Computers
Olin H. Bray and *Harvey A. Freeman*
Database Management System Anatomy
James A. Larson
Data Structures and Computer Architecture
Kenneth J. Thurber and *Peter C. Patton*
Distributed Database Management Systems
Olin H. Bray
Distributed-Processor Communications Architecture
Kenneth J. Thurber and *G.M. Masson*
Interconnection Networks for Large-Scale Parallel Processing
Howard J. Siegel
Local Networks
W.R. Franta and *Imrich Chlamtac*
Microcomputer Interfacing
G. Jack Lipovski
Requirements-Oriented Computer System Design
Kenneth J. Thurber

Basic Computer Logic

John R. Scott
Cossor Electronics, Limited

LexingtonBooks
D.C. Heath and Company
Lexington, Massachusetts
Toronto

Library of Congress Cataloging in Publication Data

Scott, John R
Basic computer logic.

1. Logic circuits. 2. Electronic digital computers—Circuits. I. Title.
TK7888.4.S26 621.381.9'585 80-5074
ISBN 0-669-03706-0

Published simultaneously in Canada

Printed in the United States of America

International Standard Book Number: 0-669-03706-0

Library of Congress Catalog Card Number: 80-5074

Contents

List of Figures

List of Tables

Preface and Acknowledgments

This book is written to provide the information required by anyone entering the digital field of electronics for the first time, whether a student or an ex-analog engineer or a technician. It is designed to provide information on the majority of digital components available, from a discussion of the most basic gates to an introduction to microprocessors. Boolean algebra, employed in digital design, is covered in chapter 3; but the book is structured in such a way that any reader wishing to avoid the algebra, and hence the design of digital circuits using it, may do so without greatly affecting his understanding of the rest of the book.

Chapter 1 provides an introduction to logic as well as a history. Logic gates are introduced by showing their construction from discrete components and hence the different logic families available on the market at the present time.

Chapter 2 introduces the basic gates with their circuit symbols and truth tables. Parameters associated with these gates are defined, including propagation delay, noise immunity, and fan out.

Chapter 3 introduces Boolean algebra and demonstrates its use in circuit design and simplification. The concepts of the algebra are utilized to show the use of Karnaugh maps for the same purpose.

Chapter 4 deals with number systems, including binary, octal, hexadecimal, Gray, BCD, and excess-3.

Chapter 5 returns to the logic circuits and shows how the basic gates introduced in chapter 2 can be connected to form bistable and monostable elements.

The construction of shift registers and counters is shown in chapter 6, and examples of synchronous counter design are considered.

Chapter 7 leads the reader through a number of types of combinational logic, including adders, comparators, multiplexers, and decoders.

No book on computer logic would be complete without the information provided in chapter 8, which details the types of memories used in computers, both past and present. Similarly, ten years ago chapter 9 would have not been included apart from its description of an arithmetic-logic unit. It introduces microprocessors, the heart of many modern computers, and their associated support circuitry.

Chapter 10 is included to show practical digital circuits and may be used by the reader for review purposes.

I am indebted to a number of people for their help in making this book possible. My thanks go the members of the Training Department of Cossor

Electronics, Limited, in particular to Mr. E.J. Killin for his encouragement and suggestions, to Mrs. J. Berry who singlehandedly deciphered my writing and typed the complete text, and to Mr. G. Stark for his assistance with the drawings. I also wish to thank my wife, Trish, for her constant encouragement and support during the writing of this book.

1 Introduction to Logic

The world we live in is an analog one, within which there are measurable quantities such as distances, rotations, weights, voltages, currents, and resistances. The last three examples may be measured using an analog meter, and in the past this has been the most common method of measuring them. The disadvantage of these meters, however, is that the reading obtained will differ according to the operator's position relative to the "needle" and thus requires interpretation by the operator. If a digital readout is presented to him instead, then this visual error is obviously overcome.

In order to make this possible, circuits that represent the values in a digital form must be available. The circuits should allow the representation of values as digits, consisting of ones and zeros (1s and 0s), these digits in turn being equivalent to high and low voltages, respectively. Thus there are now only two voltage levels instead of a voltage that is dependent on an analog value. The pattern of 1s and 0s is actually the binary number system, as will be explained in chapter 4.

This chapter will show how the digital circuits can be constructed from the basic electronic components available and, in so doing, will demonstrate, the historical development of these circuits and introduce some of the different families of circuits available. The usual name for these circuits is *logic devices.* They vary in complexity from the simple buffer up to the most complex microprocessor.

Among the simpler logic devices are those called *gates.* These gates form the basis for the more complex devices. Generally they are circuits whose inputs or outputs are defined as being either "high" or "low" and are therefore two-state devices.

In order to see how a gate may be constructed, consider the circuit of figure 1-1. Each of the inputs to the circuit may be either high (1 volt) or low (0 volts). If it is high it is said to be a logic-level 1 and, if low, a logic-level 0.

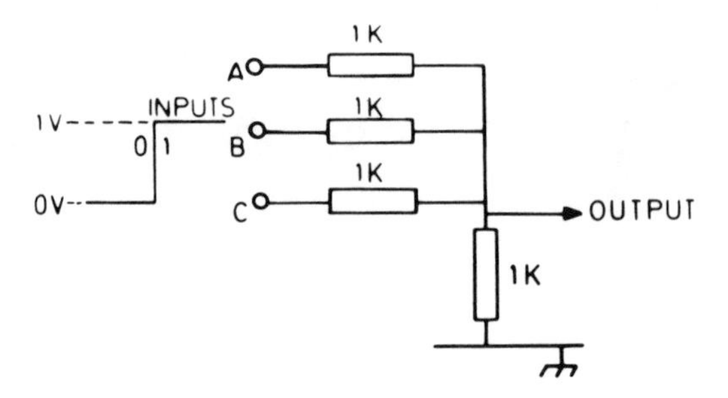

Figure 1–1. Resistor Circuit for an OR Gate

Basic resistor theory permits the voltage at the output to be calculated for all the possible input states. A table can be constructed as follows:

A	*B*	*C*	*Output Volts*
0	0	0	0
0	0	1	0.25
0	1	0	0.25
0	1	1	0.50
1	0	0	0.25
1	0	1	0.50
1	1	0	0.50
1	1	1	0.75

For example, if all inputs are 0, then obviously the output must be 0. Alternatively, if one input is a 1 and the other two are 0, then the circuit is equivalent to a 1-k resistor in series with three parallel 1-k resistors. The three resistors in parallel are equivalent to 0.33-k (applying the standard formula for resistors in parallel):

$$\frac{1}{R_{total}} = \frac{1}{R_1} + \frac{1}{R_2} + \frac{1}{R_3}$$

Therefore, the circuit is that of a potential divider with a 1-k resistor and a 0.33-k resistor in series, as shown by figure 1–2. Consequently, the output voltage V_{out} may be defined by:

$$V_{out} = \frac{R_2}{R_1 + R_2} \times V_{in}$$

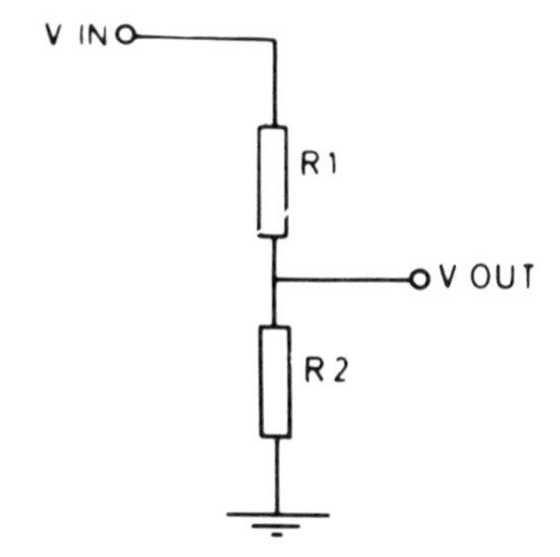

Figure 1-2. Equivalent Circuit for One Input High

Inserting the values in this formula gives:

$$V_{out} = \frac{0.33}{1 + 0.33} \times 1$$
$$= \frac{1}{4}$$
$$= 0.25 \text{ v.}$$

A similar procedure may now be carried out to complete the table for two inputs high, and then for all three inputs high. This is left to the reader as an exercise.

If the output is now defined to be a high or a 1 if it is greater than 0.2 v and a low or 0 if it is less than 0.2 v, then the output is high for any input that is high, that is, for A, B, or C being high. This circuit is therefore called the *OR gate.*

Hence a circuit that has both two-state inputs and two-state outputs has been designed. Its inherent disadvantage, however, is that there is interaction between the inputs because of the common summing resistor, and consequently there is no clearly defined output voltage for a high.

An ideal circuit for an OR gate is one that has the same function as the switch circuit of figure 1-3. For this circuit the output is a 1 when A, B, or C is a 1, that is closed. Owing to switches being used, the output is therefore always clearly defined.

This circuit is obviously impractical for the design of digital circuits; in electronics, however, a device is available that acts in a manner very similar to that of the switch: the *diode.*

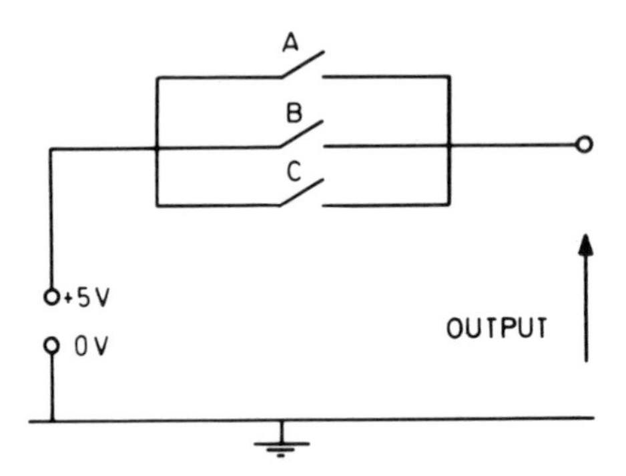

Figure 1-3. Ideal OR Gate

Positive and Negative Logic

Before we consider how diodes can be employed in place of switches, it should be pointed out that all types of logic circuits belong to one of two groups, either *positive logic* or *negative logic.*

Positive logic has already been mentioned as the type for which the more-positive signal is termed a 1 and the more-negative signal a 0, as shown in figure 1-4. It is the most common group used.

Negative logic is the inverse; that is, the more negative signal is termed a 1 and the more positive a 0, as shown in figure 1-5.

Figure 1-4. Positive Logic

Figure 1-5. Negative Logic

Diode Logic

The ratio of the back to the forward resistance for a diode is very high and in many circuits, therefore, may be regarded as an open or closed switch. Figure 1-6 shows a silicon diode when forward biased and also when reverse-biased. When forward biased, the diode typically has a voltage drop across it of 0.7 v, and therefore is not absolutely identical to a closed switch although it is a close approximation to one. A reverse-biased diode is not a

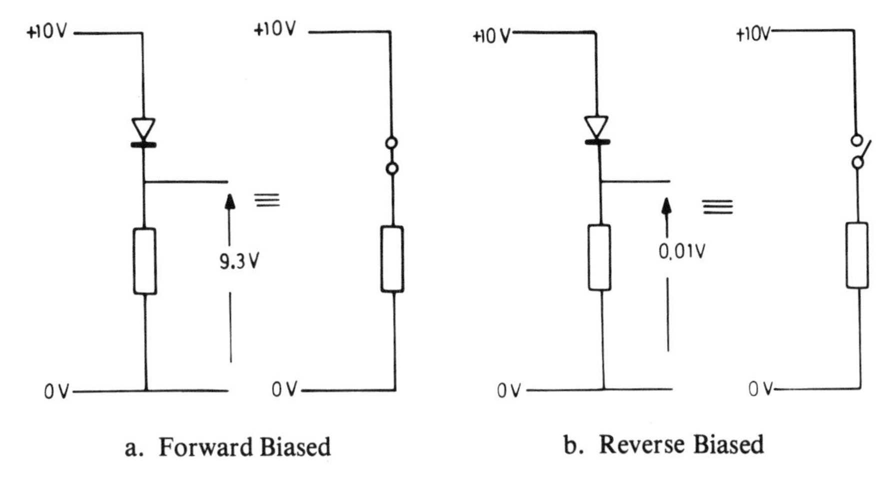

a. Forward Biased b. Reverse Biased

Figure 1-6. Forward-Biased and Reverse-Biased Diode

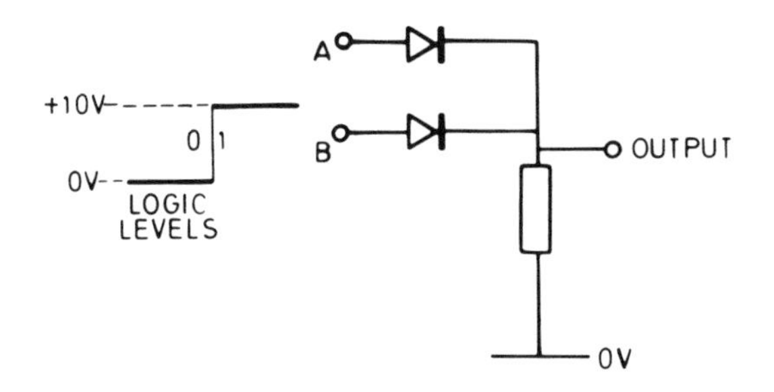

Figure 1-7. Diode Circuit for an OR Gate

perfect switch either, owing to a small current, called *leakage current,* that flows through it.

Consider the circuit in figure 1-7. The output may be defined for the different inputs as follows:

1. If both inputs (A and B) are at 0 v, then both diodes are reverse biased and the output is 0 v.
2. If both inputs are +5 v, then both diodes are forward biased and the output is therefore at +4.3 v (allowing for the forward voltage drop of the diodes).
3. If either input is at +5 v, its respective diode will conduct and the output will be +4.3 v. The diode associated with the other input will be

reverse biased, isolating that input. (Thus the problem of interaction between inputs for the resistor circuit has now been overcome.)

From items 1, 2, and 3 a voltage table may be drawn as follows:

A	*B*	*Output*
0 v	0 v	0 v
0 v	+5 v	+4.3 v
+5 v	0 v	+4.3 v
+5 v	+5 v	+4.3 v

If a logic 1 is now defined as a voltage above 2.0 v and a 0 as a voltage less than 2.0 v, then the voltages may be replaced by 0s and 1s as follows:

A	*B*	*Output*
0	0	0
0	1	1
1	0	1
1	1	1

It can be seen from this that the output is a 1 when A or B is 1. Hence the circuit is that of a diode OR gate and the table is said to be the *truth table* for the two-input OR gate. Other types of gates can also be constructed using diodes, as demonstrated in figure 1-8.

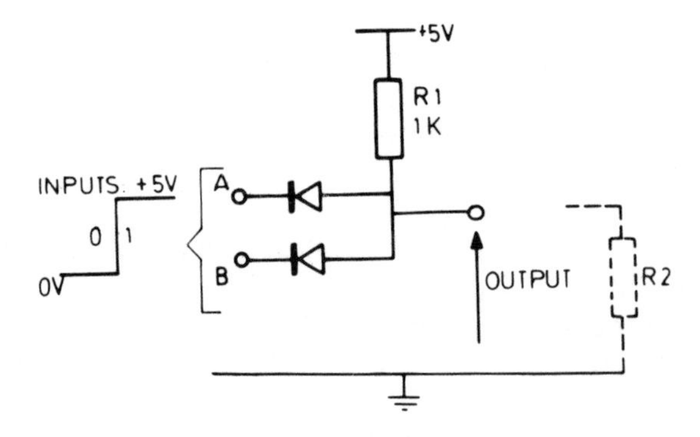

Figure 1-8. Diode Circuit for an AND Gate

For this circuit the output may be defined for the various combinations of inputs as follows:

1. If inputs A and B are both at 0 v, then both diodes will conduct and the output will be at approximately 0.7 v.
2. If both inputs are at +5 v, then both diodes are now reversed biased and the output will be a high voltage that is dependent on the load resistance R_2. For example, if R_2 is 2 k, then the output voltage will be 3.3 v.
3. Finally, when one input is at +5 v and the other at 0 v, then one diode is forward biased and the other reversed. Therefore, the output will be at 0.7 v.

Using the same definition for logic 0 and 1 given for the OR gate, 0s and 1s may be substituted for the voltages and a truth table constructed.

A	*B*	*Output*
0	0	0
0	1	0
1	0	0
1	1	1

The output is a 1 only when both A and B are 1. This is therefore a circuit and truth table for a two-input AND gate.

This section has shown how to construct an AND gate and an OR gate using diodes.

Fan In, Fan Out, and Propagation Delay

Before continuing this discussion of the construction of gates from discrete components, three terms will be defined in order to show the advantages and disadvantages of each type of logic family. These terms are *fan in, fan out,* and *propagation delay.*

First, *fan in* is defined as the number of drives that a logic element is capable of accepting. Obviously, this is partially dependent on the number of inputs the element has.

Fan out is the number of following inputs that can be driven by a single output without giving a false output.

Finally, *propagation delay* is the time necessary for a pulse applied to the input of a device to appear or cause a change at the output. It is measured at the 50-percent-of-peak points for both input and output pulses, as shown in figure 1-9.

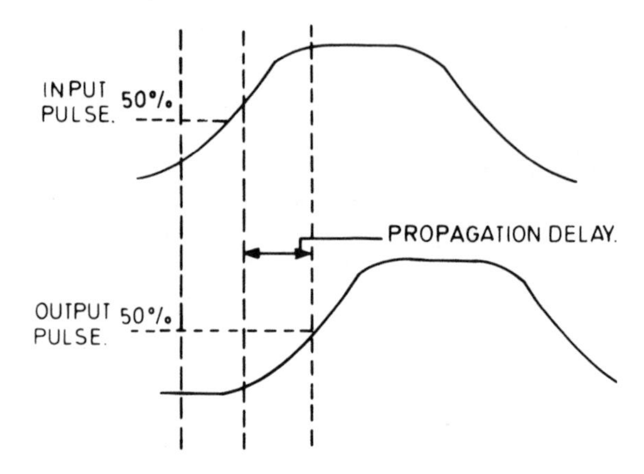

Figure 1-9. Propagation Delay

Disadvantages of Diode Logic

Unfortunately, diode gates such as those in figures 1-7 and 1-8 have a number of disadvantages, even more than do most other logic families. These can be summarized as follows:

1. There is a voltage attenuation owing to the forward-voltage drop across the diodes of 0.7 v. As increasing numbers of gates are connected together, the attenuation becomes so large that the output is always a 0.
2. Because of this voltage attenuation, the range of resistance values becomes further restricted with each additional level of gating.
3. The switching time, and therefore the propagation delay, is increased with each additional gate owing to load and stray capacitances.
4. Finally, the fan out of diode logic is limited by the lack of available output current, and the fan in is limited by the diode's leakage current.

Resistor-Transistor Logic (RTL)

The disadvantages of using diodes as switches was demonstrated in the previous section. The next development is the use of a transistor as a switch. Transistors overcome the problem of dc-level degradation experienced by diodes, and they may be "cut off" to a high-impedance state or "turned on," so that they have a very low resistance. When working in this mode the transistors are outside their normal amplifying region and are either cut off or saturated, as demonstrated in figure 1-10 (points A and B). This diagram

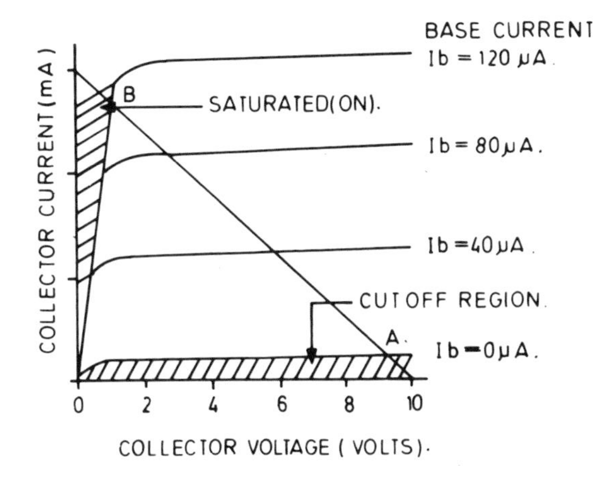

Figure 1-10. Typical npn Transistor Output Characteristics

shows a typical output characteristic of an npn transistor in common emitter configuration.

It can be seen from figure 1-10 that when the base current is nil, then the npn transistor is cut off (point A) and the collector voltage is high. Also, when the base current is high the transistor will be on (point B), and the collector voltage is low.

Therefore, for an npn transistor, if the base potential is raised above that of the emitter by more than 0.7 v, then the transistor will conduct. If the base potential is lowered, then the transistor is nonconducting or off. This is demonstrated in figures 1-11a and 1-11b. The arrows indicate a voltage change in the positive or negative directions. Similar rules can be

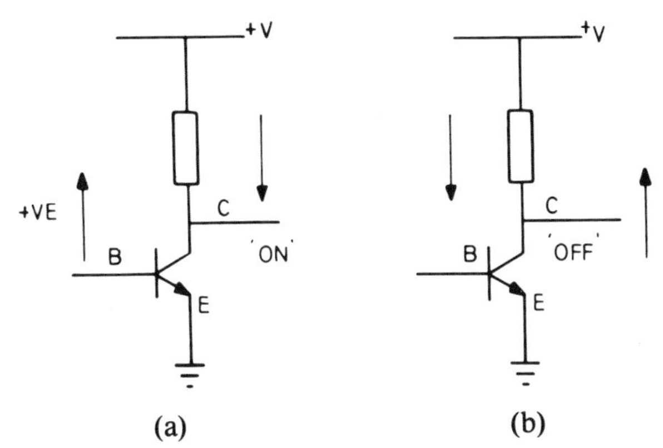

Figure 1-11. npn Transistors Used as a Switch

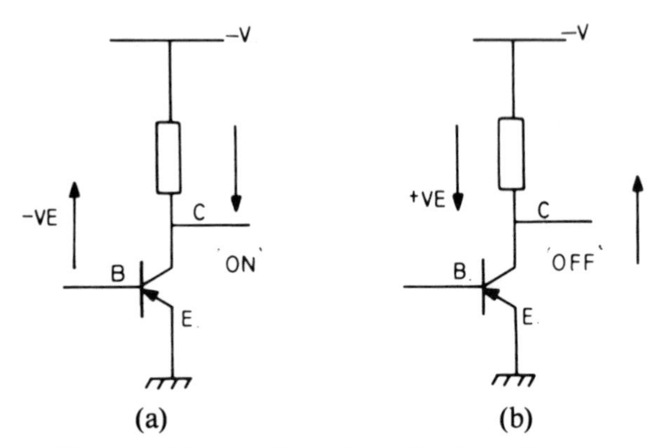

Figure 1-12. pnp Transistor Used as a Switch

applied to pnp transistors except that a more positive potential on the base now turns it off and a more negative potential turns it on. This is demonstrated in figures 1-12a and 1-12b.

If a potential divider network is now connected to the base of the transistor and an input signal is applied to the top resistor of the network, as shown in figure 1-13, then it is possible to switch the transistor on and off.

If the input is at 0 v—a logic 0—then the base of the transistor will be negative, holding it off, and its collector will therefore be high, a logic 1. Alternatively, if the input is at +5 v (1), then the base will be positive, switching the transistor on, causing the collector to be low. Hence the truth table for the circuit is:

Input	*Output*
0	1
1	0

The circuit inverts the input and is therefore called an inverter or a NOT gate.

This circuit has been constructed using resistors on the input and a transistor output and it is therefore called *resistor-transistor logic* (RTL). This type of logic was in fact the first type to be introduced in the now more-common integrated-circuit form of logic.

In practice it used a +3.8-v power supply instead of +5 v. A typical RTL circuit for a logic gate is given in figure 1-14. For this circuit the output is low whenever one of the transistors is conducting, that is, when

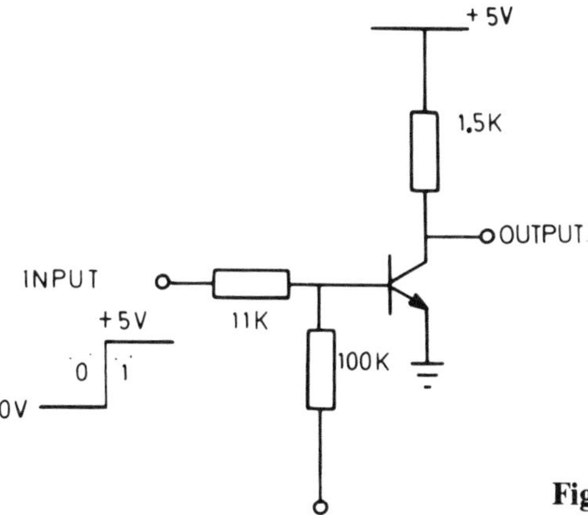

Figure 1-13. Resistor-Transistor Inverter

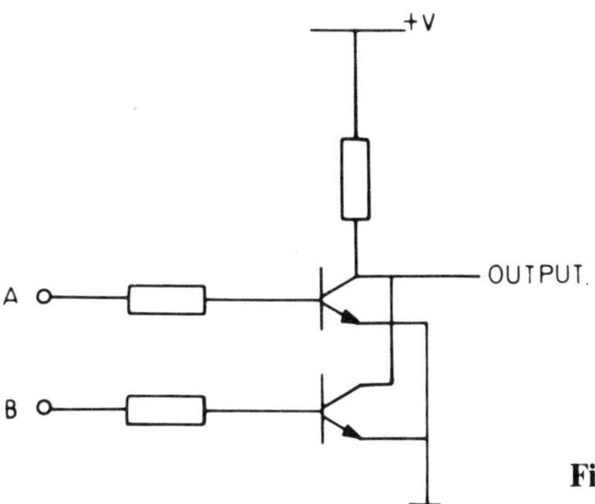

Figure 1-14. Typical RTL Circuit for a NOR Gate

any input is high. Therefore, the output is high only when both inputs are low; and the truth table for this circuit is:

A	*B*	*Output*
0	0	1
0	1	0
1	0	0
1	1	0

Referring back to the truth table for the OR gate, it may be seen that this output is actually the inverse of that for the OR gate. It is therefore called a NOT OR gate or, more commonly, a NOR gate.

RTL has a typical propagation delay of 12 nsec, a fan out of 4, and a power dissipation of between 30 and 100 mW per gate. This type of logic was used widely in the early 1960s, but it has been rendered obsolete by types developed since that require less space.

Diode-Transistor Logic (DTL)

This was the next logic family developed after RTL. Initially the circuits were constructed from discrete components, but they were very soon integrated. As the name suggests, DTL gates are constructed from both diodes and transistors: diode inputs and a transistor output. The transistor overcomes the problem of voltage attenuation experienced with diode logic.

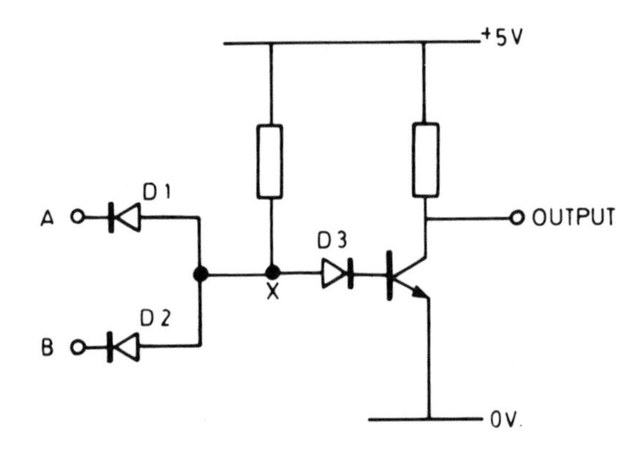

Figure 1-15. A Typical DTL Gate

Figure 1-15 is an example of a typical DTL gate.
For the circuit of figure 1-15:

1. If either input is low, then point X will be low and D3 will be reverse biased, causing the transistor to be cut off and its output to be high. (Diode D3 acts as a level-shifting diode and ensures that the transistor is off under these input conditions. If it were not present, then the current would be shared between the grounded input diode and the base of the transistor).
2. If both inputs are high, then both D1 and D2 will be reverse biased and

point X will be high also. Hence D3 is forward biased and the transistor is switched on, causing the output to be low.

For DTL a logic 0 and a logic 1 are defined as 0.4 v and 3.5 v, respectively. Therefore, using items 1 and 2, the truth table for the circuit is:

A	*B*	*Output*
0	0	1
0	1	1
1	0	1
1	1	0

The truth table is actually the inverse of that for an AND gate, and the gate is consequently a NOT AND or NAND gate.

As for diode logic, the fan in of DTL is limited. It is also slower (it has a longer propagation delay of about 30 nsec than transistor-transistor logic (TTL), which has been developed since.

Transistor-Transistor Logic (TTL)

Transistor-transistor logic is really a more advanced version of DTL. It replaces the input diodes of DTL with transistors connected in common-base configuration. TTL has a lower propagation delay than DTL, typically 10 nsec., and has become the most popular logic family in use today. Figure 1-16 demonstrates how a gate may be constructed using transistors on the inputs instead of diodes.

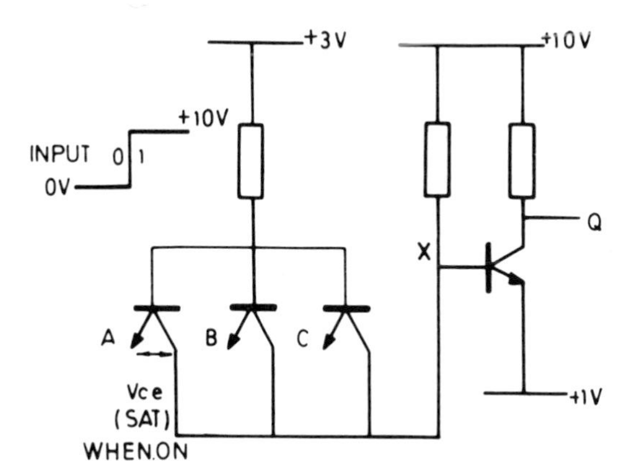

Figure 1-16. Transistor-Transistor NAND Gate

If all three inputs A, B, and C are high, then all three transistors will be cut off. Therefore, the base of the output transistor will be high, the transistor will be on, and hence Q will be low. However, if any input is low, then the corresponding input transistor will saturate and point X will fall to a potential equal to V_{ce} (saturated). The output transistor will therefore be turned off (its emitter has been raised to +1 v to exceed V_{ce} (sat) to ensure this) and the output will be high. Q is thus low only when all inputs are high, and consequently this is a circuit for a three-input NAND gate.

Integrated-Circuit TTL

Using integrated-circuit techniques it is possible to form multi-emitter transistors and therefore to replace the three transistors given in figure 1–16 with one three-emitter transistor. This has an advantage over the previous circuit because a greater uniformity of characteristic voltages may be obtained. It has the added advantage of reducing input capacitance, thus reducing switching time. Figure 1–17 shows a two-input TTL NAND gate.

Within the TTL family of devices there are a number of ranges of TTL with different characteristics. These will be discussed in detail in chapter 2.

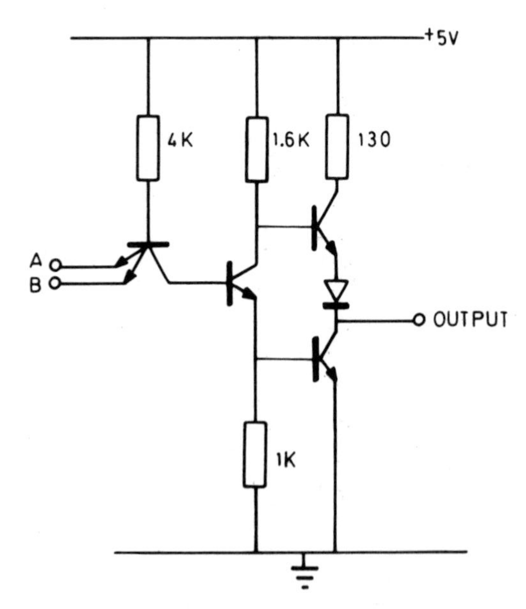

Figure 1–17. Two-Input TTL NAND Gate

Integrated-Injection Logic (I^2L)

This type of logic is one of the most recent introductions to the bipolar-logic range. Like TTL, it employs transistors working in the saturated region of their characteristics.

I^2L gates may contain both npn and pnp transistors in the same circuit, such as that shown in figure 1-18. This gives I^2L an advantage, as the collector of TR1 and the base of TR3 are the same p-type material and can therefore be the same physical element, thus increasing the packing density of the circuits over that of TTL. The packing density is also increased because there are no resistors in the circuit. I^2L has an added advantage in being slightly faster than standard TTL, having a propagation delay of 5 nsec.

In figure 1-18 TR1 and TR2 act as current sources to the bases of TR3 and TR4. If input A or input B is high, then TR3 or TR4 will conduct and the output will be low. However, if both inputs are low, then the bases of TR3 and TR4 will be effectively shorted to ground, holding the transistors off and causing the output to be high. This circuit is therefore that of a NOR gate.

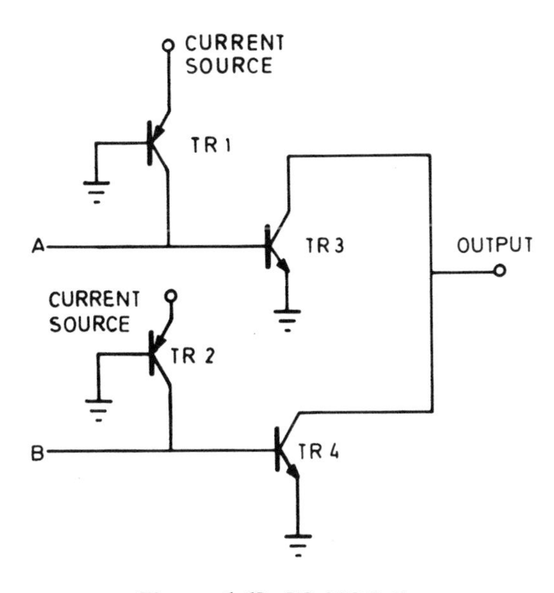

Figure 1-18. I^2L NOR Gate

Emitter-Coupled Logic (ECL)

All the transistor-logic families introduced so far use transistors working in the saturated mode. This has a disadvantage because a capacitance is built up within the transistor when it saturates, and this capacitance will consequently reduce the switching speed. Unlike these types of logic, ECL is non-saturating and is therefore faster. It can achieve switching rates of up to 1,000 MHz, compared to 35 MHz for standard TTL, and has a propagation delay of only 1 or 2 nsec. Thus it is the fastest of all present-day logic circuits. Among its characteristics are a high input impedance and a low output impedance, and hence a high fan out of 16. It also provides simultaneous complementary outputs without using external inverters.

The disadvantages of ECL are that it consumes a moderate amount of power (30 mW per gate) and is not particularly immune to noise.

ECL requires a −5.2-v supply, and typical high and low output voltages are −0.96 v and −1.65 v, respectively. Figure 1-19 is a typical ECL circuit.

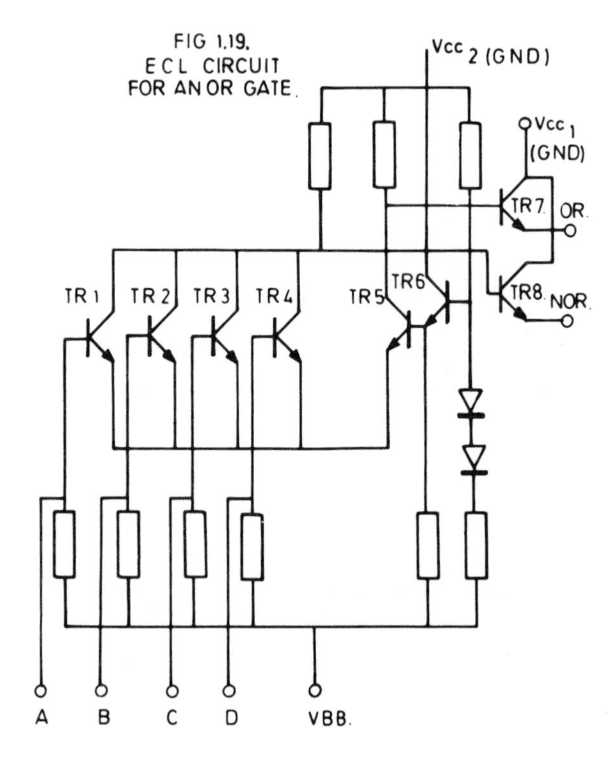

Figure 1-19. ECL Circuit for an OR Gate

Metal-Oxide-Semiconductor Logic (MOS)

In the late 1960s and early 1970s a new type of semiconductor device was developed: the *field-effect transistor* (FET). The manufacturing technique for this device uses metal-oxide-semiconducting materials, and thus the MOSFET is a metal-oxide-semiconductor field-effect transistor. MOSFETs can be constructed in two complementary ways, as either PMOS or NMOS, the latter being used extensively now in the manufacture of microprocessors. [P stands for P doping and N stands for N doping where doping is the introduction of impurities into the semiconductor in order to make the dominant charge carriers electrons (N-type) or holes (P-type)]. Figure 1–20 shows the circuit symbols for the FETs.

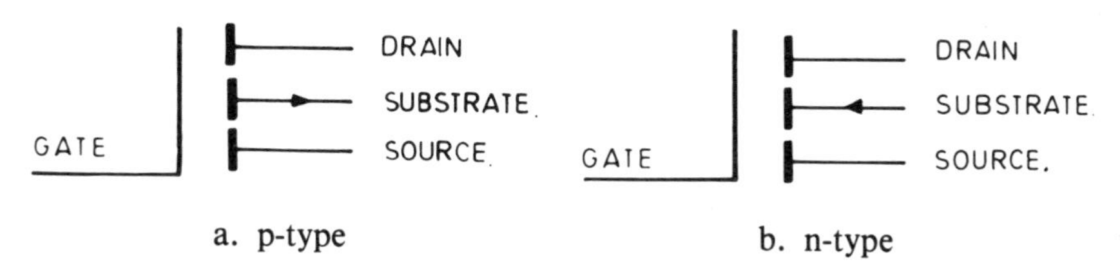

Figure 1–20. Circuit Symbol for a FET

The switching of FETs may be considered to be similar to that for bipolar transistors. Both types of FET have a threshold voltage associated with them, which determines whether or not the transistor is conducting. For an n-channel transistor the voltage applied to the gate must be greater than V_{TN} (where V_{TN} = n-channel threshold voltage) for it to conduct. If it is less than V_{TN}, then the transistor is off. For a p-channel FET the voltage applied to the gate must be less than $V_{DD} - V_{TP}$ (where V_{DD} = supply volts and V_{TP} = p-channel threshold volts) for the transistor to be conducting. If it is greater than this value, then it is off.

Another logic family was developed soon after the FET. This combined both the PMOS and the NMOS techniques in a complementary manner to produce a more-ideal switching action than standard PMOS. Because of its structure it was therefore called *complementary metal-oxide-semiconductor logic* or CMOS. Figure 1–21 shows a schematic for a CMOS inverter together with a plot of its switching characteristics compared with that of PMOS. Figure 1–22 shows a schematic for a CMOS two-input NOR gate. With both inputs low (less than V_{TN}), TR1 and TR2 are conducting and TR3 and TR4 are cut off, causing the output to be high. If input 1 is high, the TR3 will be conducting, TR1 will be cut off, and the output will be low.

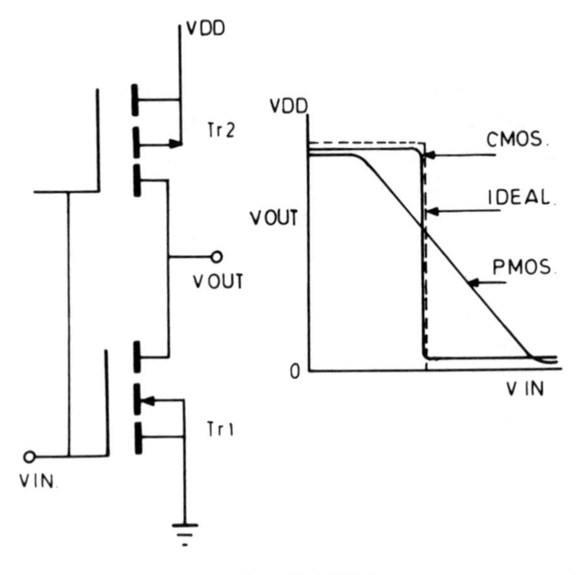

Figure 1-21. CMOS Inverter

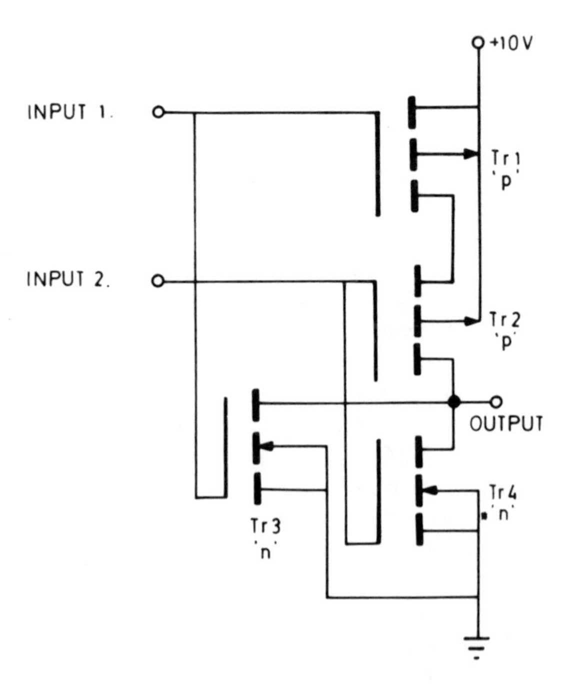

Figure 1-22. CMOS Two-Input NOR Gate

Similarly, if input 2 is high, TR4 will conduct and Q1 will be cut off so that the output will also be low. Hence the output must also be low for both inputs high, and the circuit is consequently that of a two-input NOR gate.

The advantages of CMOS are that it has a lower power dissipation (10 nW per gate), because of the extremely high off resistance of MOS transistors, and a high noise immunity owing to its complementary configurations. It also has a high fan out of 50 and is relatively high speed, although not quite as fast as standard TTL, with propagation delays of between 12 and 60 nsec depending on the supply voltage used, which in turn may be between 3 and 15 volts. Additionally, it has a high packing density, the highest of all types and much better than that of TTL. This enables large-scale integrated circuits to be produced, such as microprocessors, memories, and other microprocessor-support devices.

Exercise 1.1

1. Explain the difference between positive and negative logic.
2. Give two disadvantages of diode logic.
3. Define *fan out* and *propagation delay.*
4. What active device may be used to overcome the dc-level degradation present in diode logic?
5. What are RTL, DTL, TTL, and ECL? Which of these is the fastest?
6. What are the main advantages of MOS?

2 Basic Logic Gates

Chapter 1 gave a brief introduction to logic gates by showing how some of the basic gates may be constructed from discrete components. Although the gates are actually constructed of these components, they are not drawn on circuit diagrams in this form since this would be too cumbersome. There are specific symbols to represent each of the basic logic functions: AND, NAND, OR, NOR, NOT, EXCLUSIVE OR, EXCLUSIVE NOR, and the BUFFER. The descriptions of these functions that follow include details of truth tables and circuit symbols, as well as examples of the gates constructed in TTL.

The AND Gate

The function of any gate is uniquely defined by its truth table. This defines what the output of the gate will be for a specific set of inputs. The AND-gate truth table is as follows:

Inputs		*Output*
A	B	F
0	0	0
0	1	0
1	0	1

It can be seen from this table that the output of an AND gate is a high (logic-level 1) only when both A *and* B inputs are high.

An example of a two-input TTL AND gate is given in figure 2-1. This shows part of the circuitry that is actually in the AND-gate integrated-circuit package. In practice there are normally four two-input AND gates within the package, and therefore the amount of circuitry shown in figure 2-1 must be quadrupled.

Integrated-circuit versions of the AND gate may have two, three, four, or more inputs. This is achieved by using multi-emitter transistors, as described in chapter 1.

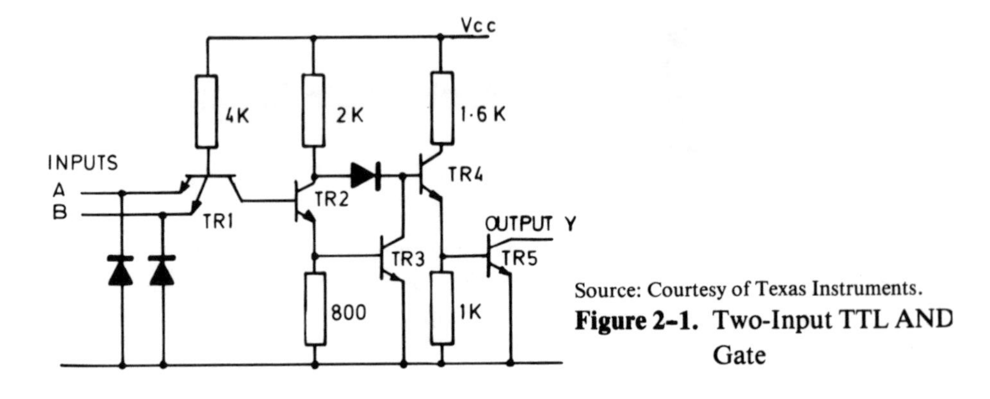

Source: Courtesy of Texas Instruments.

Figure 2-1. Two-Input TTL AND Gate

A number of different sets of circuit symbols are used to define logic gates. Depending on the set used, there is always one symbol within the set that uniquely defines the AND gate. The symbols used in this book are the commonly used American symbols, although the British symbols are shown later in the chapter.

The logic symbol for the AND gate is:

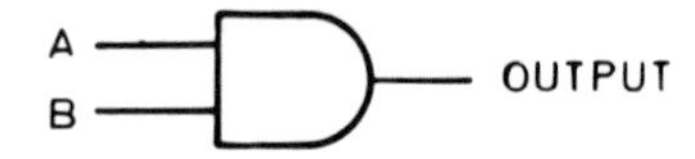

When designing or fault finding a circuit it is often useful to draw out the circuit signals as they appear relative to time—in fact, just as they would appear if viewed on an oscilloscope. Such a drawing is called a *timing diagram.* The following is a typical example of such a diagram, drawn to demonstrate the AND function.

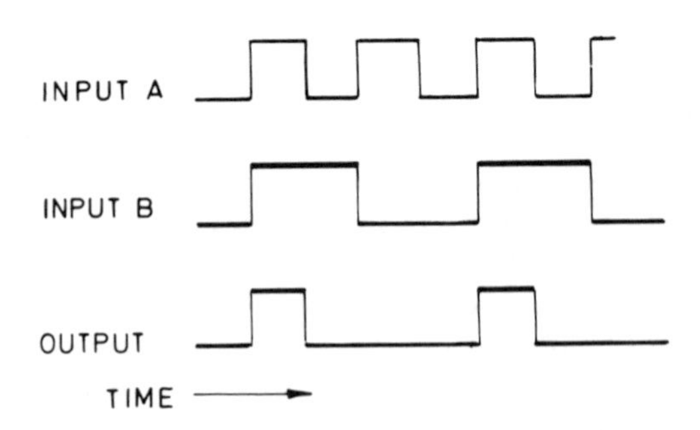

A very simple example of the use of an AND gate is in an airplane seat-belt monitor, where all the seat belts must be "made" to give an indication to the pilot.

The OR Gate

This type of gate gives a high output when one input or another input is a high or when more than one input is high. This is demonstrated by the truth table for the OR gate, which is as follows:

Inputs		*Output*
A	B	F
0	0	0
0	1	1
1	0	1
1	1	1

The logic symbol for the OR gate is:

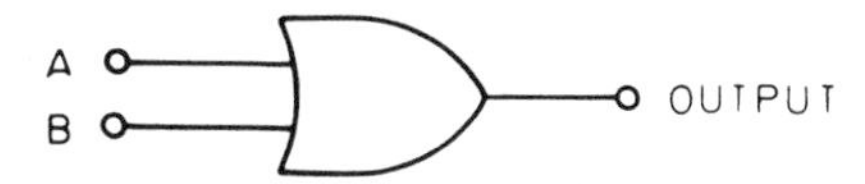

Once again a timing diagram can be drawn to demonstrate the OR-gate function. In this example, as with that of the AND gate, the signal applied to the A input is twice the frequency of that applied to the B input; and therefore the timing diagram is:

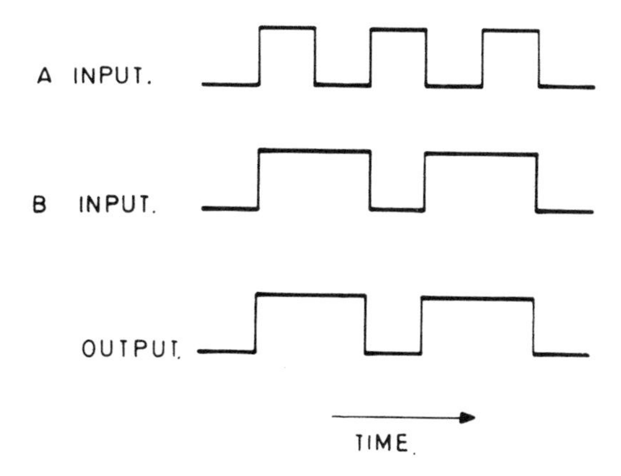

The NOT Gate

This gate differs from the AND and OR gates in that it has only one input. Its output state is always the opposite of its input state, and it is therefore normally called an inverter. Thus the truth table for it is:

A Input	*Output*
0	1
1	0

The logic symbol for the NOT gate or *inverter* is:

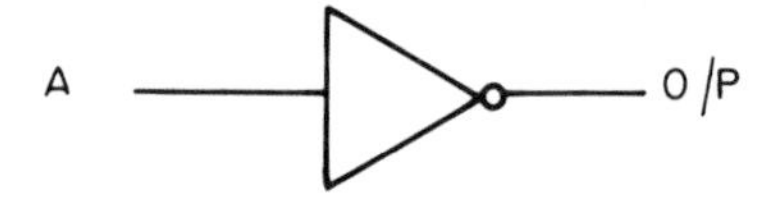

This is the most commonly used symbol, although sometimes the small circle (indicating inversion) may appear on the input line rather than the output.

"Highs" or "logic 1s" and "lows" or "logic 0s" have been considered, but so far exactly what they are has not been stated, other than that a 1 is a high voltage and a 0 is nominally 0 v. In fact, for standard TTL a 1 is typically a voltage greater than +2.4 v and a 0 a voltage less than +0.4 v. This subject will be dealt with in more detail later in this chapter.

The inverter, together with all the gates considered so far, has only two output states: a 1 or a 0. However, there is a device called a *tri-state inverter* that has a third output state: that of high impedance. This state occurs when when a second input line called the *enable* line goes to a logic 0. Its logic symbol and its truth table are as follows:

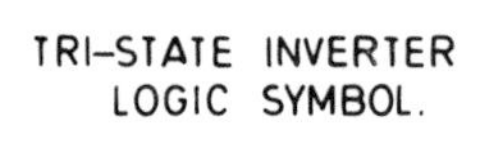

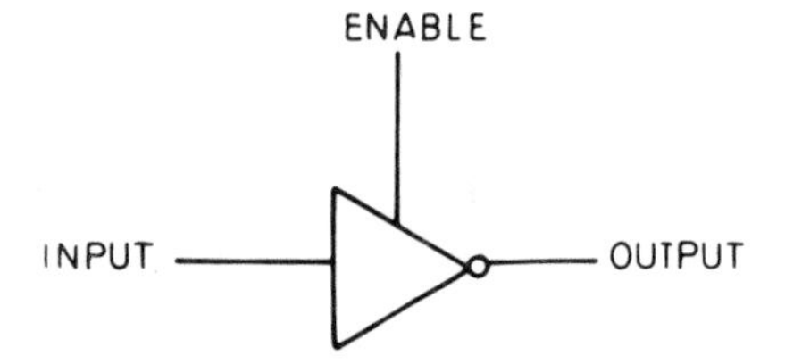

Input	*Enable*	*Output*
0	1	0
1	1	0
X	0	Z

X = Don't care (0 or 1).

Z = High impedance.

The Buffer

The output of a *buffer* is the same as its input. At first glance there may seem to be very little application for such a circuit, as a length of wire performs the same function. However, practical circuits are often amplifiers or isolators, which isolate one electronic circuit from another.

An example of a buffer is a transistor connected to the output of a low-power logic circuit to enable it to turn on a lamp. The circuit cannot power the lamp on its own, since the current flowing through the lamp (and therefore through the circuit) would eventually overheat and possibly damage or even destroy the delicate circuit. A suitable transistor, however, can easily dissipate the excess heat and drive the lamp when it is switched on by the logic circuit.

The logic symbol for this gate is:

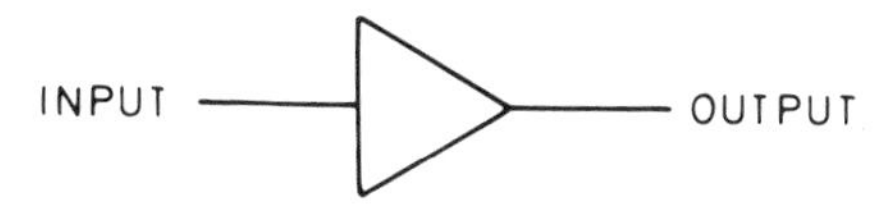

As with the tri-state inverter, there is also a tri-state version of the buffer. This has an enable line as well, and when it is low the output is in the high-impedance state as demonstrated by the truth table.

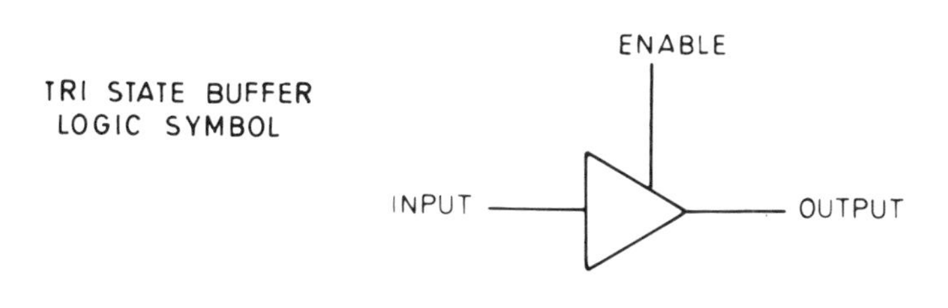

Input	*Enable*	*Output*
0	1	0
1	1	
X	0	Z

X = Don't care (0 or 1)

Z = High-impedance state

Both tri-state inverters and buffers are used extensively in computers for isolating one part of an electronic circuit from another.

The NAND Gate

This type of gate produces a low output when both A *and* B inputs are high as shown in the truth table.

Inputs		*Output*
A	B	F
0	0	1
0	1	1
1	0	1
1	1	0

Thus it is the complement or negation of the AND gate and may be considered to be an AND gate and an inverter, as shown in the following diagram:

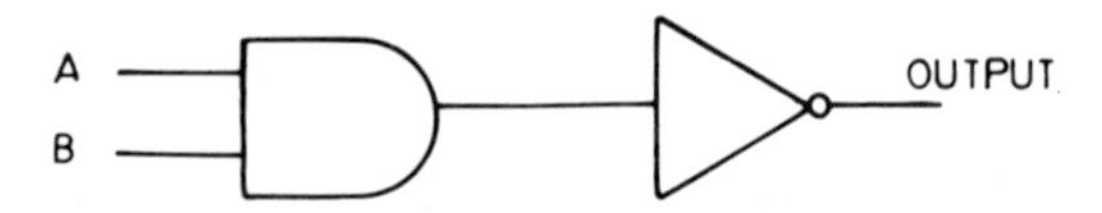

Rather than perform this function in a circuit, in this way an integrated-circuit package containing NAND gates would be used and when drawn on the circuit diagram would have the following logic symbol:

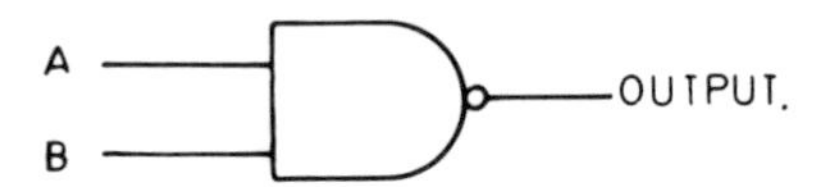

A timing diagram for a two-input NAND gate is as follows:

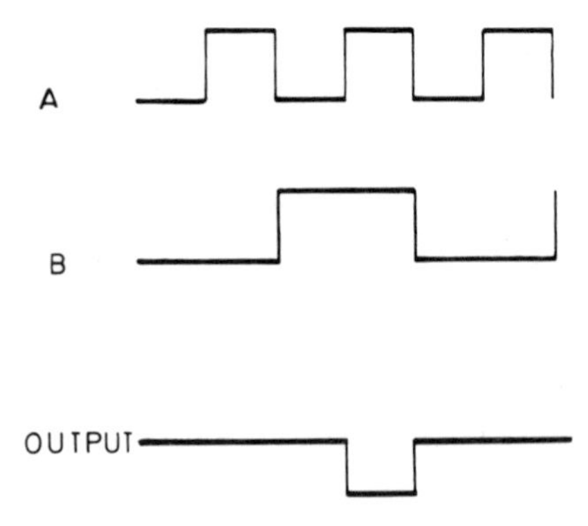

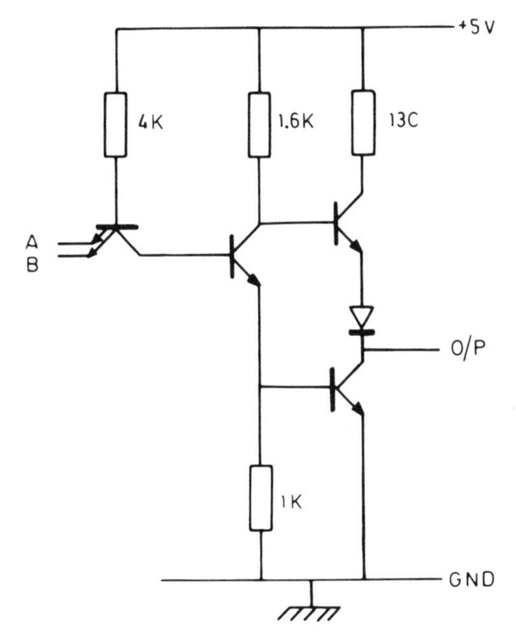

Source: Courtesy of Texas Instruments.
Figure 2–2. Two-Input TTL NAND Gate

A typical example of how a NAND gate may be constructed in TTL is demonstrated by figure 2–2.

The NOR Gate

The NOR function is the complement of the OR function and as such is effectively an OR gate with an inverter connected to its output:

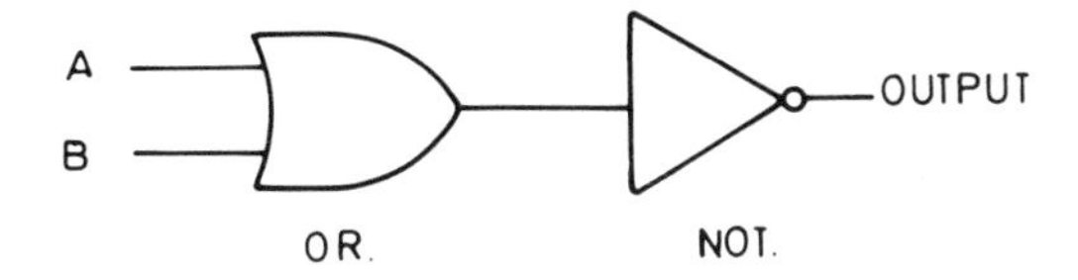

As with the NAND gate, there is one symbol to represent the function rather than the two shown in the diagram. The logic symbol for it is:

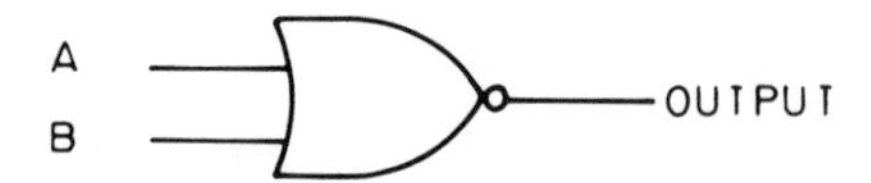

The output of a NOR gate is therefore a high when A *or* B is low, as indicated by the truth table. The truth table for a NOR gate is:

Inputs		*Output*
A	B	F
0	0	1
0	1	0
1	0	0
1	1	0

A timing diagram for this gate is of the form:

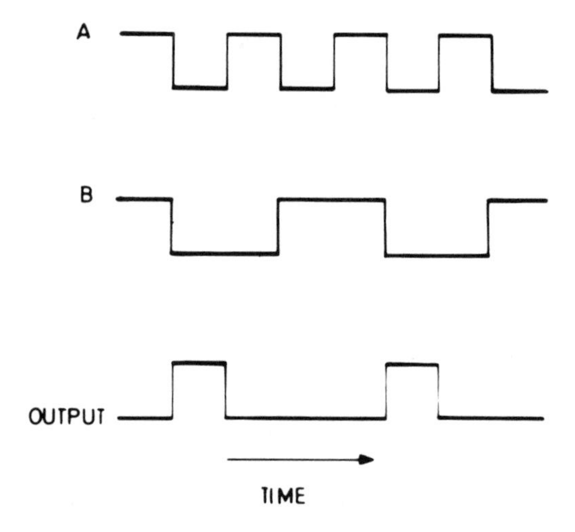

A NOR gate constructed from TTL is shown in figure 2-3.

Obviously, a large number of the gates considered are connected together to perform specific functions; and it is therefore important to be able to work through a particular circuit to find its output. Where the inputs of such a circuit are constantly changing with respect to time, a timing diagram must be drawn to find the output.

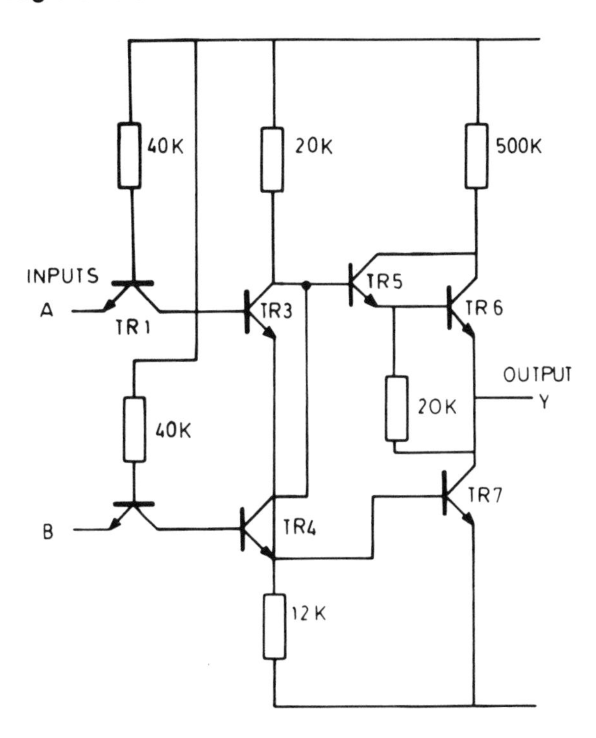

Source: Courtesy of Texas Instrument.

Figure 2-3. Two-Input TTL NOR Gate

Example 2.1

Find the output Q of the following circuit for the input waveforms given.

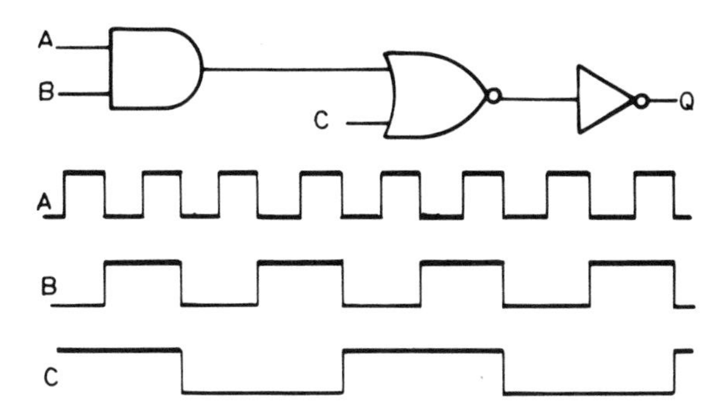

Exercise 2.1

Find the output Q of the following circuit for the input waveforms shown.

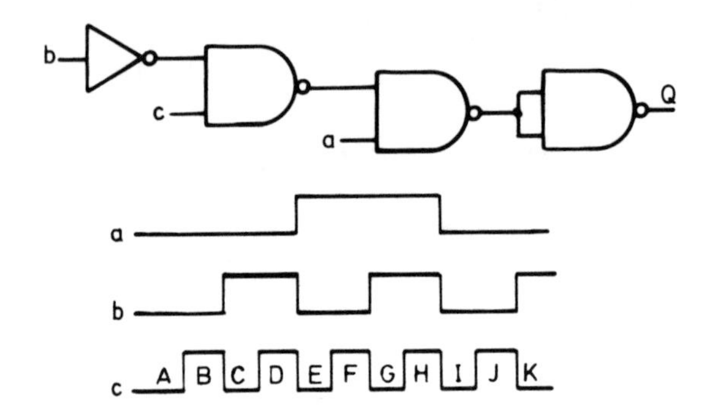

Exercise 2.2

Find the output Q of the following circuit.

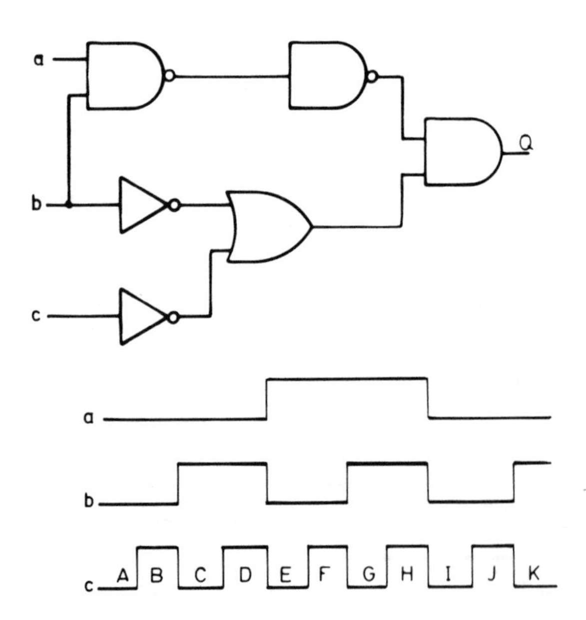

The EXCLUSIVE OR Gate

The truth table for the EXCLUSIVE OR gate is:

Inputs		*Output*
A	B	F
0	0	0
0	1	1
1	0	1
1	1	0

That is, the output of an EXCLUSIVE OR gate is high when the inputs A and B differ.

The logic symbol for it is:

A
B
OUTPUT

We can draw a timing diagram for the gate as follows:

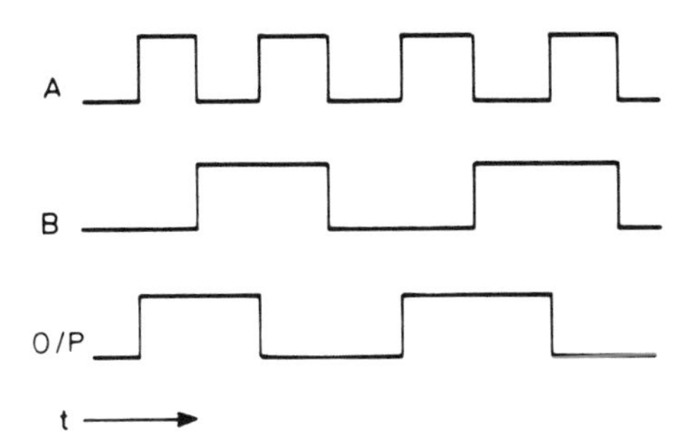

An application of an EXCLUSIVE OR gate is the comparator. It can be seen from the truth table that it will give an output when two inputs fail to compare. This property may be used to check the parity of a binary number. (Note that the parity of a binary number is an indication of whether there are an even number of 1s in the number or an odd number of 1s, and is used extensively in computer memory banks as a corruption check and also when data is being transferred from one computer to another.)

The EXCLUSIVE NOR Gate

The truth table for an EXCLUSIVE NOR gate is:

Inputs		*Output*
A	B	F
0	0	1
0	1	0
1	0	0
1	1	1

That is, the output of an EXCLUSIVE NOR gate is high when the inputs A and B are the same.

The logic symbol for it is:

A
B
OUTPUT

We can draw a timing diagram for the gate as follows:

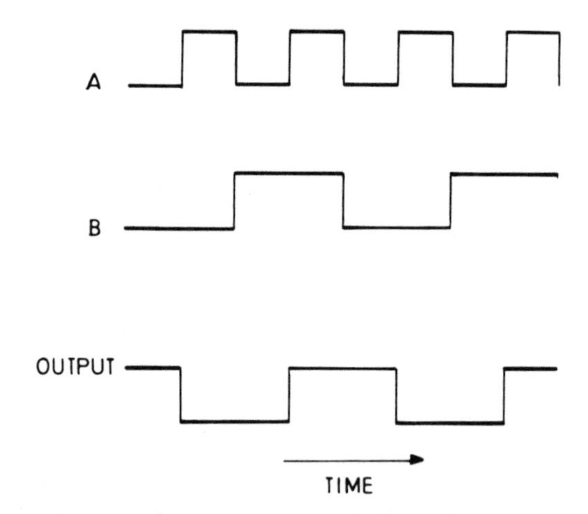

An application of an EXCLUSIVE NOR gate is as an equality detector, the gate producing an output when both inputs are the same.

Equivalent Circuits

Although the basic logic functions considered so far are normally drawn on circuit diagrams using the basic logic symbols, they may also be drawn differently.

Example 2.2

To find the equivalent gate to the following circuit:

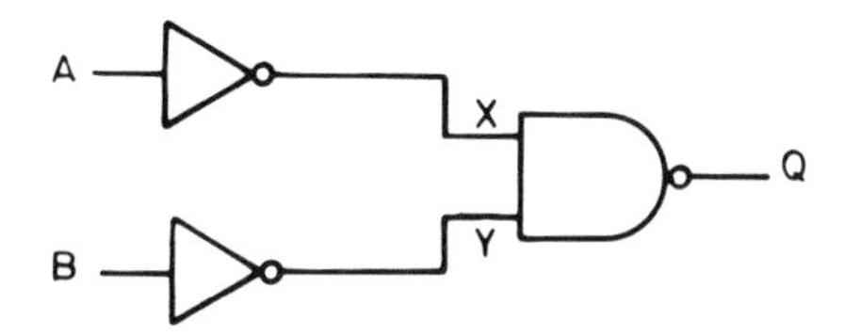

A truth table may be constructed step by step as follows:

Inputs		*Outputs*		
A	B	X	Y	Q
0	0	1	1	0
0	1	1	0	1
1	0	0	1	1
1	1	0	0	1

Observe inputs A and B and output Q. The output is high when A *or* B is high. Therefore, the overall function performed by the circuit is that of an OR gate, and it could have been drawn as one gate only. Also, one gate would have been sufficient to perform the function rather than three, thereby saving on integrated-circuit packages. Therefore:

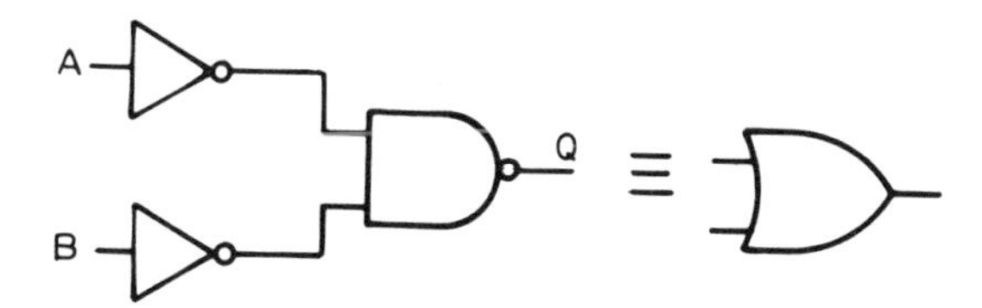

An alternative method of drawing the gate is with small circles on the inputs, as follows:

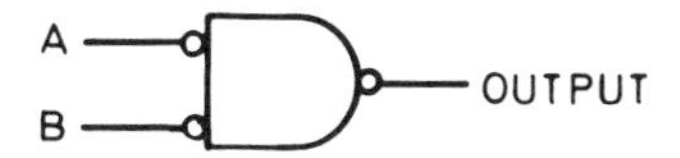

This indicates that a logic 0 on input A *and* a logic 0 on input B produces a logic 0 at the output, and that at no other time would the output be 0, that is, an OR gate.

Exercise 2.3

Find the equivalent gates of the following:

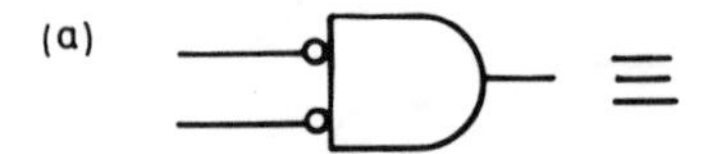

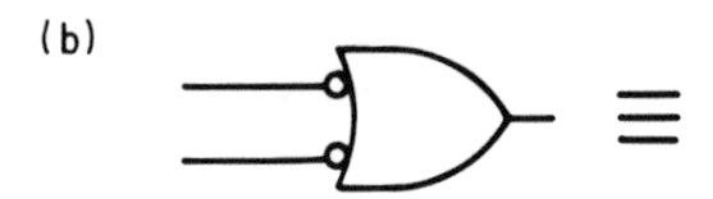

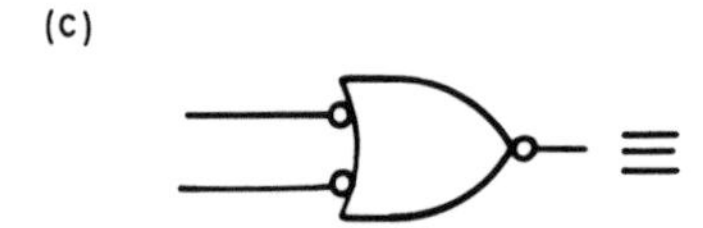

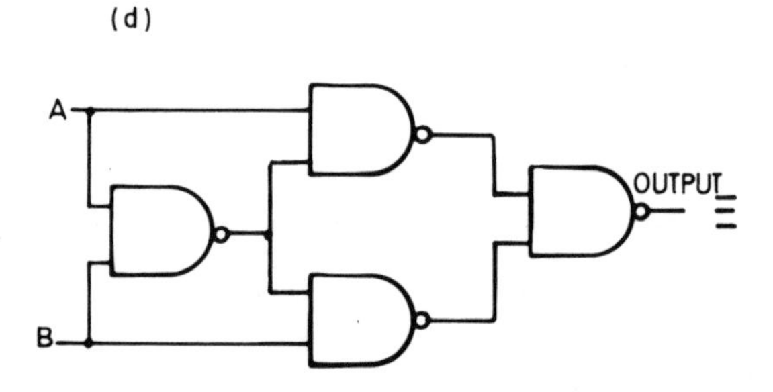

Open-Collector and Totem-Pole Outputs

Consider the TTL circuits previously drawn for the AND, NAND, OR, and NOR gates. The AND-gate output differed from those of the other gates in that the example given was an open-collector output whereas the other examples were of totem-pole outputs.

Open-collector devices have the advantage that several of the devices may be wired together to form a multiple AND gate using only one resistor as follows:

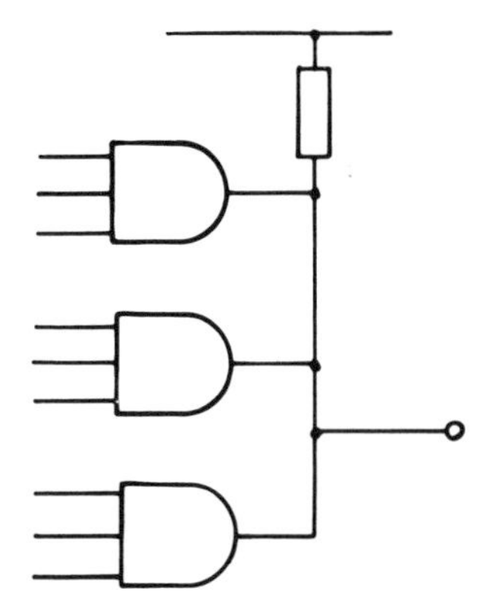

Another advantage of open-collector devices is that the current drawn by the output stage and with it the fan out can be adjusted so that the stage is not working at maximum fan-out capability where there may be noise problems.

With the totem-pole type of output, the "upper" transistor is employed as a load, which acts as a high resistance when off and as a low resistance for low output impedance when on. Figures 2–4 and 2–5 show the currents and voltages in both output states.

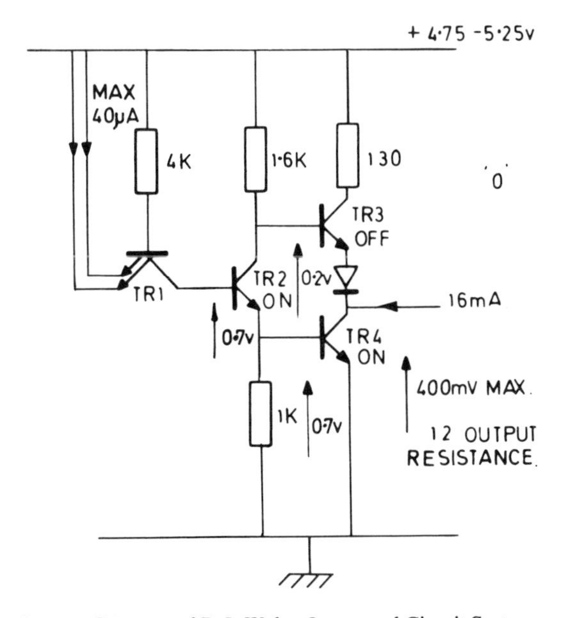

Source: Courtesy of D.J. Walter Integrated Circuit Systems.

Figure 2–4. 0 = Output-State TTL NAND Gate

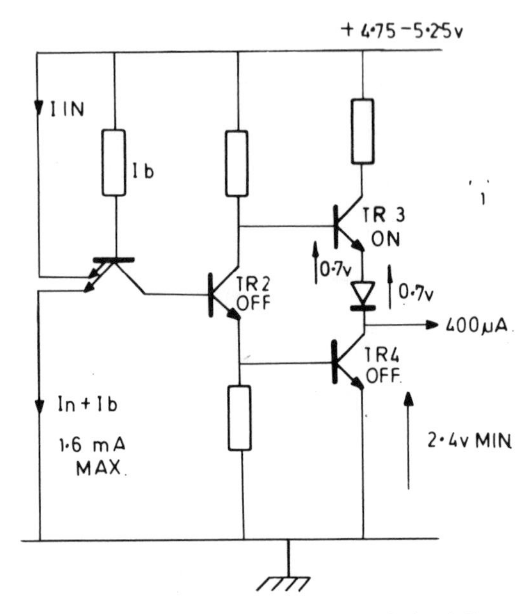

Source: Courtesy of D.J. Walter Integrated Circuit Systems.

Figure 2-5. 1 = Output-State TTL NAND Gate

The output transistor TR4 can "sink" a maximum current of 16 mA and therefore can accept current from ten inputs at 1.6 mA each. TR4 is also capable of providing a sourcing current of 400 μA, which allows it to drive ten inputs, each taking 40 μA. Thus TTL has a maximum fan out of ten regardless of whether the output is a 0 or a 1. The number of emitters on a gate can vary from device to device, and this therefore limits the fan in.

Noise Margins and Threshold Levels

Any signal that is not intended to be at a particular point in a circuit is *noise*. Noise may be caused by a relay, by crosstalk from other gates, by spikes on the supply rail caused by switching or coupling from main inputs, or by reflections caused by driving long lines. It can cause a circuit to give a false output; and therefore the susceptibility of a system to noise, or its noise immunity, is very important.

One measure of the susceptibility of a circuit to noise is the *noise margin* for the gates within it. This noise margin is defined as the difference between the output voltage and the input threshold. The input threshold for a TTL gate is the input voltage at which the output will change state; it has a typical value of 1.4 v. The output voltages for a gate in a 0 state and a 1 state are approximately 0.2 v and 3.3 v, respectively. Hence the noise margin for a gate in the 1 output state = 3.3 − 1.4 = 1.9 v, and in the 0 output state = 1.4 v − 0.2 v = 1.2 v.

However, the manufacturer generally guarantees the output to be less than 0.4 v in the logic-0 state and greater than 2.4 v in the logic-1 state, and also generally guarantees that the threshold will lie between logic 0.8 v and 2.0 v. Consequently, the guaranteed noise margin for worst-case devices is (0.8 − 0.4) and (2.4 − 2.0), that is, for both states, 400 mV. Hence for the worst case, a TTL gate can tolerate 400 mV of dc noise before giving a false output.

Unused Inputs

Degradation in switching time and noise problems may occur if unused inputs are left unconnected. To avoid such problems, therefore, unused inputs should be terminated using one of the following methods:

1. Unused inputs are connected to used ones, if this does not exceed the driving capability of the driving output, that is, if the fan out is not exceeded.
2. Connect unused inputs to supply via a 1-k pull-up resistor. This protects the inputs from transients. A maximum of 25 inputs may be connected to a 1-kΩ resistor.
2. Connect unused inputs to the output of an unused gate, ensuring that the output of the unused gate is always high.
4. Connect unused inputs to separate supply volts between 2.4 v and 3 v.

It is likely that of these four methods the first two are the most commonly used.

Present Ranges of TTL

Apart from the standard range of TTL, low-power, high-speed, Schottky, and low-power Schottky ranges are also available. All have the same basic circuit configuration and are compatible with each other in such things as supply and logic voltages, but there is a compromise between speed and power. In order to achieve higher speed and lower propagation delays, the circuit resistor values have been reduced. This in turn reduces all the time constants that result from base capacitances, stored charge, and stray and load capacitances, thereby giving a faster propagation of the signal through the gate. Unfortunately, reducing the resistor values causes a higher power consumption. This means that the product of power and speed for a family is approximately constant. Thus if one plots low power, standard TTL, and high speed, as in figure 2-6, they lie on the same hyperbolic curve.

Adding Schottky clamp diodes, as shown in figure 2-7, improves the speed of the family without increasing power, so that the two Schottky families lie on a better curve. The improvement in speed is achieved because the diodes have a forward voltage that is lower than the voltage required to forward bias the collector-base diode of a silicon transistor, thus preventing saturation, which leads to reduced switching time. The different powers and speeds of the families are shown in table 2-1.

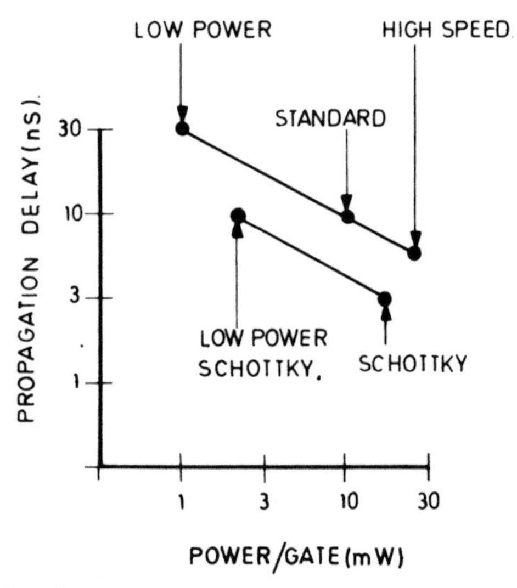

Source: Courtesy of Texas Instruments.

Figure 2-6. Power-Speed Compromise of the Various Types of TTL Circuits

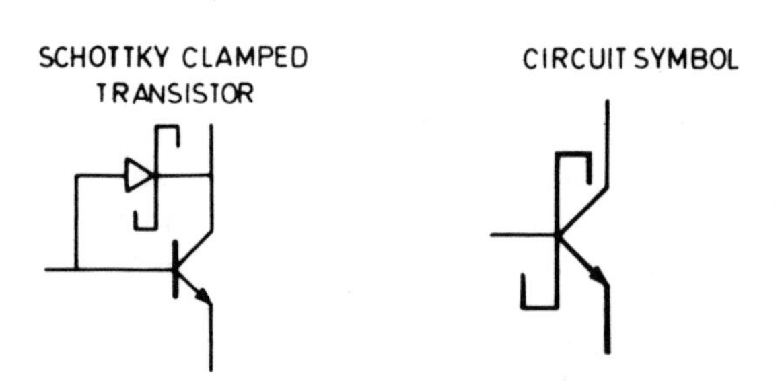

Figure 2-7. The Schottky Clamped Transistor

Table 2-1
TTL Family Comparison

Type of TTL	*Average Gate Propagation Delay (nS)*	*Average Power per Gate (mW)*	*Speed-Power Product (pJ)*
Low power	33	1	33
Low-power Schottky	10	2	20
Standard	10	10	100
High-speed	6	23	138
Schottky	3	19	57

Schmitt Trigger Inputs

As previously stated, the typical threshold level for TTL is 1.4 v. However, within the standard TTL range there are gates available that have Schmitt trigger inputs. This causes a hysteresis effect because the threshold level for a positive transition of the input is different from that for a negative transition. Figure 2-8 demonstrates this for a two-input NAND gate.

Pulse shaping and simple resistor-capacitor (RC) multivibrators are among this particular circuit's applications. Figure 2-9 shows a simple multivibrator that can form the basis of a wide-frequency-range clock pulse for clocking logic circuits. The circuit is self-starting, and by changing the value of capacitor C it is possible to alter the frequency of operation.

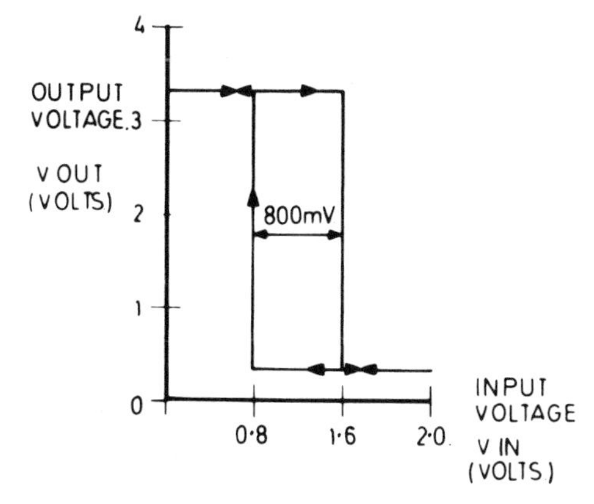

Figure 2-8. V_{in} $-V_{out}$ Characteristic of Schmitt Trigger NAND Gate Showing Hysteresis

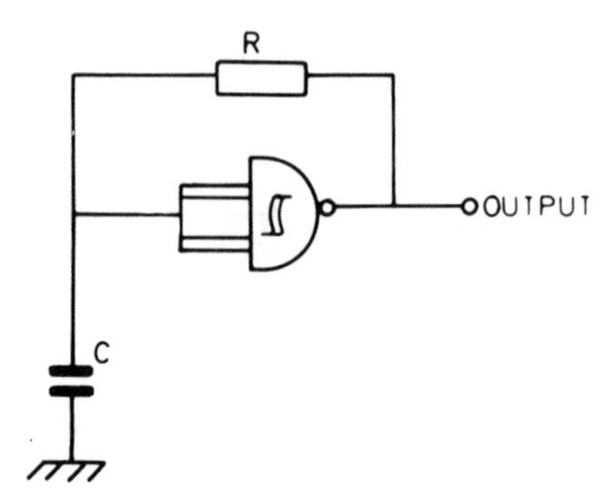

◀ **Figure 2–9.** RC Multivibrator

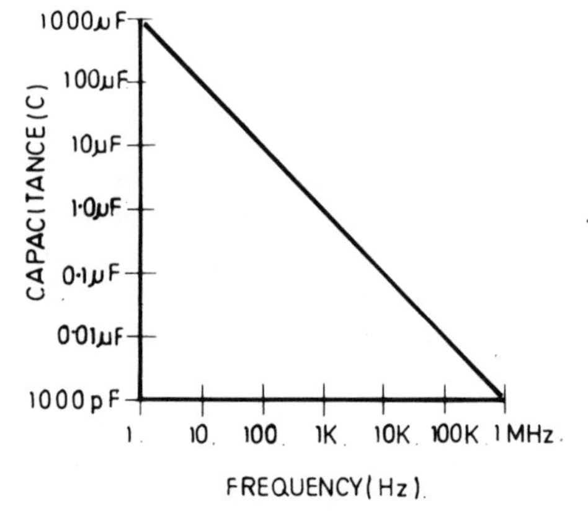

Source: Courtesy of Texas Instruments.

◀ **Figure 2–10.** Frequency Versus Capacitance for R = 390Ω

Initially, capacitor C is discharged; thus the NAND gate output is logical 1. When power is applied to the circuit, the capacitor charges up through resistor R until the upper threshold voltage is reached. The output then changes to a logical 0, and capacitor C discharges to the lower threshold voltage through resistor R. The cycle is then repeated.

The voltage dropped across resistor R determines its limiting values when the input is approaching the lower threshold and the output is at a logical 0. Figure 2–10 is a graph of pulse-repetition frequency versus values of capacitor C.

The Data Sheet

There are an enormous number of logic devices on the market, and when designing a circuit it is important to know the specifications to which each device is manufactured. This information can be found in the manufac-

CIRCUIT TYPES SN5400, SN7400
QUADRUPLE 2-INPUT POSITIVE NAND GATES (1)

schematic (each gate)

W FLAT PACKAGE (TOP VIEW)

4Y 4B 4A GND 3B 3A 3Y
14 13 12 11 10 9 8
1 2 3 4 5 6 7
1A 1B 1Y V_{CC} 2Y 2A 2B

J OR N DUAL-IN-LINE PACKAGE (TOP VIEW)

V_{CC} 4B 4A 4Y 3B 3A 3Y
14 13 12 11 10 9 8
1 2 3 4 5 6 7
1A 1B 1Y 2A 2B 2Y GND

(2)

NOTE: Component values shown are nominal.

positive logic: $Y = \overline{AB}$

recommended operating conditions

	MIN	NOM	MAX	UNIT	
Supply Voltage V_{CC}: SN5400 Circuits	4.5	5	5.5	V	(3) (4)
SN7400 Circuits	4.75	5	5.25	V	(5)
Normalized Fan-Out From Each Output, N			10		(6)
Operating Free-Air Temperature Range, T_A: SN5400 Circuits	−55	25	125	°C	(7)
SN7400 Circuits	0	25	70	°C	(8)

electrical characteristics over recommended operating free-air temperature (unless otherwise noted) (9)

PARAMETER		TEST FIGURE	TEST CONDITIONS†			MIN	TYP‡	MAX	UNIT	
$V_{in(1)}$	Logical 1 input voltage required at both input terminals to ensure logical 0 level at output	1				2			V	(10)
$V_{in(0)}$	Logical 0 input voltage required at either input terminal to ensure logical 1 level at output	2						0.8	V	(11)
$V_{out(1)}$	Logical 1 output voltage	2	V_{CC} = MIN, I_{load} = −400 μA	V_{in} = 0.8 V,		2.4	3.3		V	(12)
$V_{out(0)}$	Logical 0 output voltage	1	V_{CC} = MIN, I_{sink} = 16 mA	V_{in} = 2 V,			0.22	0.4	V	(13)
$I_{in(0)}$	Logical 0 level input current (each input)	3	V_{CC} = MAX,	V_{in} = 0.4 V				−1.6	mA	(14)
$I_{in(1)}$	Logical 1 level input current (each input)	4	V_{CC} = MAX,	V_{in} = 2.4 V				40	μA	(15)
			V_{CC} = MAX,	V_{in} = 5.5 V				1	mA	(16)
I_{OS}	Short-circuit output current§	5	V_{CC} = MAX		SN5400	−20		−55	mA	(17)
					SN7400	−18		−55		(18)
$I_{CC(0)}$	Logical 0 level supply current	6	V_{CC} = MAX,	V_{in} = 5 V			12	22	mA	(19)
$I_{CC(1)}$	Logical 1 level supply current	6	V_{CC} = MAX,	V_{in} = 0			4	8	mA	(20)

switching characteristics, V_{CC} = 5 V, T_A = 25°C, N = 10

PARAMETER		TEST FIGURE	TEST CONDITIONS		MIN	TYP	MAX	UNIT	
t_{pd0}	Propagation delay time to logical 0 level	65	C_L = 15 pF,	R_L = 400 Ω		7	15	ns	(21)
t_{pd1}	Propagation delay time to logical 1 level	65	C_L = 15 pF,	R_L = 400 Ω		11	22	ns	(22)

† For conditions shown as MIN or MAX, use the appropriate value specified under recommended operating conditions for the applicable device type.
‡ All typical values are at V_{CC} = 5 V, T_A = 25°C.
§ Not more than one output should be shorted at a time.

Source: Courtesy of Texas Instruments.

Figure 2–11. Data Sheet for TTL Circuit Types SN5400 and SN7400

turer's data sheet for the device that is given in the manufacturer's data book. This section gives an example of an extract from a typical Texas Instruments data sheet for a Quad two-input NAND package (figure 2–11), together with a key describing each part of it.

Symbolization (*Device Recognition*)

SN54H102N. This is a typical TTL-device symbolization. It can be divided into distinct parts, each of which tells us something about the device.

SN/54/H/102/N

SN. This is the standard prefix for a *semiconductor network.*. There are variations such as RSN:, BL:, and SNX:, which indicate a *radiation-hardened circuit,* a *beam-lead-constructed device,* or an *experimental circuit,* respectively.

54. TTL is available to meet three temperature ranges. Prefixes 54, 64, and 74 distinguish them.

Series 54 = −55 to +125°C military

Series 64 = −40 to +85°C

Series 74 = 0 to +70°C industrial

There is also a difference in the supply voltage range.

Series 54 = 4.5 v to 5.5 v.

Series 64 and 74 = 4.75 v to 5.25 v

H. Indicates a high-speed device. The letter(s) might have been:

L: low power

S: Schottky

LS: low-power Schottky

or there might have been no letter, which would indicate a standard-family device.

102. The next two or three numbers show the device function (102 = JK flip-flop.

N. This letter is the package type. Eleven possibilities are shown in the data book, but N is the most widely used. N = 14, 16, or 24 pin dual-in-line plastic.

SN54H102N. It should now be apparent that the example just given is of a high-speed JK flip-flop meeting the military temperature range.

This information on the data sheet was supplied by courtesy of Texas Instruments.

Absolute-Maximum Ratings

Supply voltage: V_{CC} 7.0 v

Input voltage: V_{in} 5.5 v

Electrical breakdown and irreversible damage may occur if the V_{CC} is raised above 7.0 v. This does not imply that one gets correct logical operation for all voltages below 7.0 v.

The input voltage with respect to ground (or the most-negative input) must not be greater than 5.5 v. The input will tolerate a maximum current into the gate of about 2 mA. Thus where inputs do not go to an output but to the supply rail, a current-limiting resistor should be included if it is possible that there may be transients on the rail, or negative undershoots on other inputs to that input transistor.

Basic Data Sheet

A line-by-line examination of the data sheet for the basic gate shown in figure 2–11 will show what relevance the parameters have for the user.

1. *Title.* Data sheets are usually for both Series 54 and Series 74 devices. Apart from pin connections, data-sheet parameters are the same regardless of the package.
2. *Schematic.* Resistor values are nominal and may vary by ±20 percent. *Pin configuration.* Note that the pin connections for flat package are not necessarily the same as dual in line.
3. Although the devices will operate outside the stated limits of voltage temperature, some of the parameters may then be outside the data-sheet limits.

4 & 5. Voltage ranges over which the series are specified. Series 54 is the same as Series 74.

6. Fan out is the number of standard loads (standard inputs) that the circuit outputs will drive correctly, that is, with full noise margin.

7 & 8. Temperature ranges over which the series are specified. Series 54 is specified for the temperature range from −40 to +85°C. In all other respects, however, it is identical to Series 74.

9. The table following is of typical and worst-case characteristics.

10. & 13. When the input voltage is greater than 2.0 v, the output will be less than 0.4 v even when sinking 16 mA, and even when V_{CC} is the minimum applicable value, as in 4 and 5.

11 & 12. When the input voltage is less than 0.8 v, the output will be greater than 2.4 v even when sourcing 400 μA, and even when V_{CC} is the minimum applicable value, as in 4 and 5.

14. When the input is taken down to 0.4 v, the input will source a maximum of 1.6 mA even at the maximum supply. This is because of current flowing from the V_{CC} rail via the base resistor of the input transistor, as in figure 2–12.

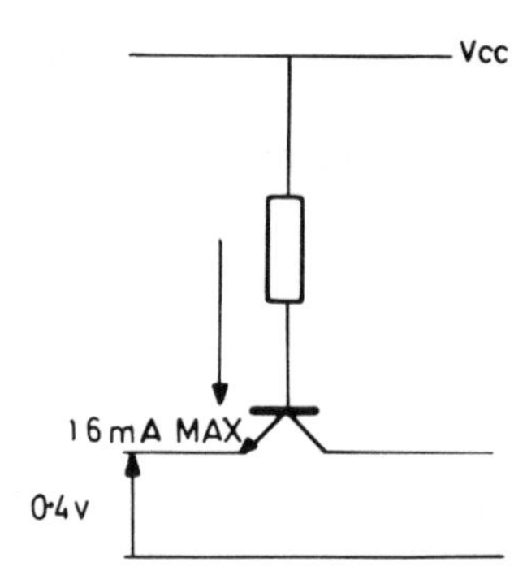

Figure 2–12. 0.4-v Input to TTL Gate

15. When the input is taken up to 2.4 v it will sink up to 40 μA even when the supply is the maximum. (14 and 15 define the standard load.)

16. If the input is taken to 5.5 v, it will sink a maximum of 1 mA even at the maximum supply. This is a breakdown condition.

17 & 18. If the output terminal is taken down to ground when the gate output is in the logical-1 state, this is the current that will flow out of the output terminal.

19 & 20. This is the total current taken by the package at typical and worst-case supply and inputs.

21 & 22. These are propagation delays measured using a circuit as in figure 2–13 to simulate ten standard loads.

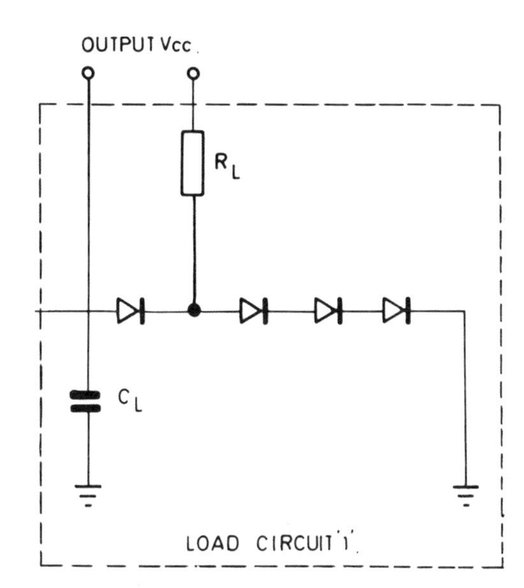

Figure 2–13. Load Circuit

Circuit-Complexity Classification

There are basically four different classifications of circuit complexity for integrated circuits: small-scale, medium-scale, large-scale, and very-large-scale integration.

Very-large-scale integration (VLSI) is defined as a system that is fabricated as a single microcircuit that contains 1,000 or more gates, or a circuit of similar complexity. The other classifications are defined in the same way except that *large-scale integration* (LSI) is defined as a system of 100 or more equivalent gates, *medium-scale integration* (MSI) as 12 or more equivalent gates, and *small-scale integration* (SSI) as one of less complexity than the others.

The gates considered in this chapter and demonstrated by the data sheet are all examples of small-scale integration. Such devices as counters and decoders are medium scale, and microprocessors are examples of large-scale or even of very-large-scale integration.

Alternative Logic Qualifying Symbols

Although the logic symbols in use throughout this book are correct, they are actually the American Standard symbols. Other standards, such as the

British Standard, vary slightly. There are also some companies that use their own particular symbols.

In general, the other symbols are variations on those already shown with either an & symbol in them, signifying the AND function, or a 1, signifying the OR. Figure 2–14 shows examples of the British Standard symbols and the American Standard symbols.

Exercise 2.4

1. What are the two types of output stages for TTL?
2. Define the noise margin.
3. State typical voltage values for logic-0 state.
4. State two methods of terminating unused inputs. Why are these procedures carried out?
5. Name the five TTL families. What is the result of adding Schottky clamp diodes in TTL circuits?
6. What type of integration is a Quad two-input NAND package?
7. Complete the following table defining the output state for varying input conditions on the gates considered in this chapter.

Inputs				*Outputs*				
EXCLUSIVE							EXCLUSIVE	EXCLUSIVE
A	B	AND	OR	NOT	NAND	NOR	OR	NOR
0	0	0						
0	1	0						
1	0	0						
1	1	1						

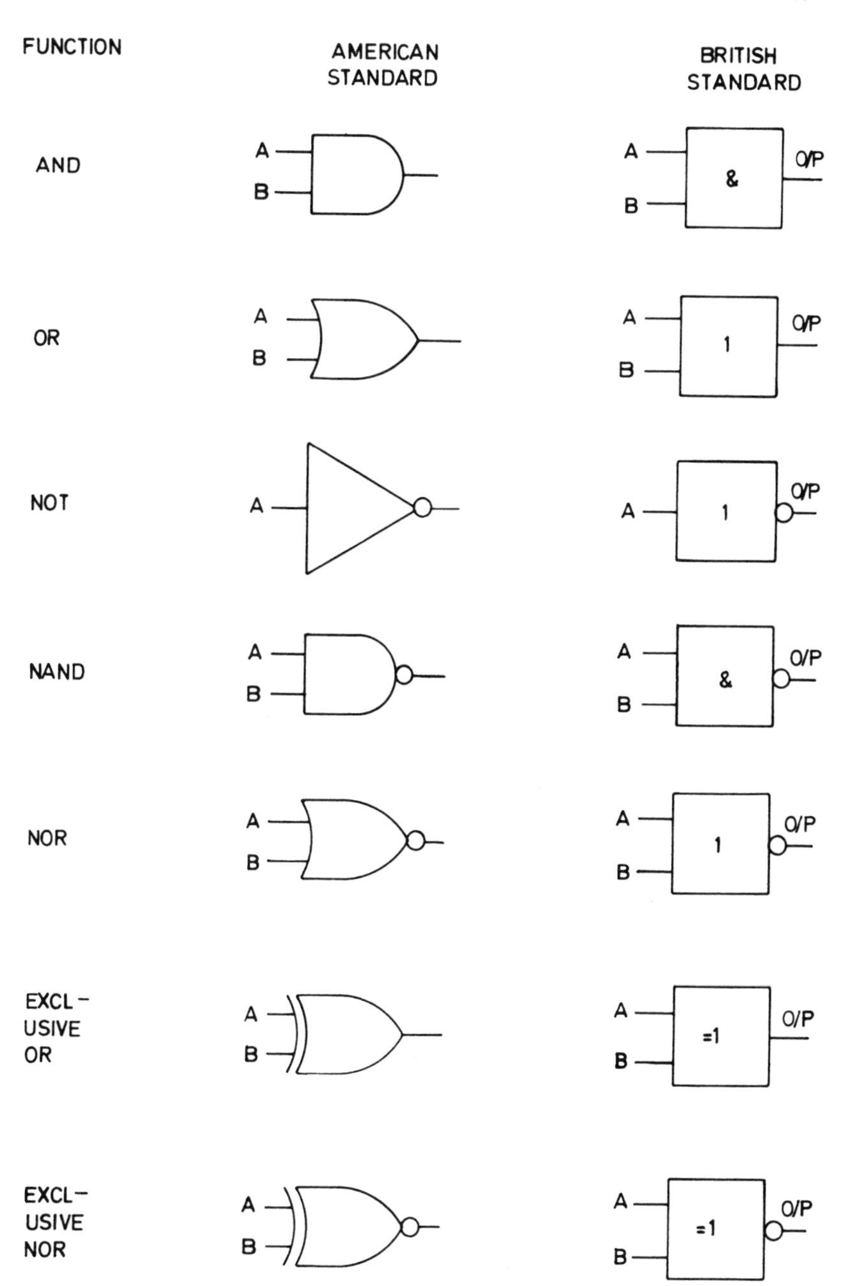

Figure 2–14. Symbols Used for Logic Gates

3 Boolean Algebra

Mathematics is a kind of shorthand that enables us to present a physical process on paper, with symbols that may be manipulated in order to gain a better understanding of the process. It is therefore a tool that aids understanding.

The familiar kind of algebra that deals with variables x and y, and makes up combinations such as $x + y$, $x - y$, $x \cdot y$, y/y, and x , is a linear process, because the two variables can hold any value. This is the kind of algebra that is performed by analog operational amplifiers.

However, if x and y have only two possible states, as in the case of a voltage that is present or absent, we can ignore the actual value of the voltage and regard the variables as behaving according to a two-stage or binary number system. The two states can be high or low and positive or negative.

Another type of mathematical algebra is therefore required to cope with this situation. For the origin of this algebra we must go back to 1854, when George Boole developed a mathematical system of logic appropriately named Boolean algebra.

Not until 1938, when Claude E. Shannon wrote a paper entitled "A Symbolic Analysis of Relay Switching Circuits," was the potential of Boolean algebra fully realized. The rules of this type of algebra which are somewhat different from those of linear algebra, will be explained in this chapter.

By using Boolean algebra it is possible to design a circuit to perform a specific function with the minimum number of logic gates. However, the minimum number of gates is not always the best solution to the problem; the number of packages used and the cost must also be considered. The cost of an extra gate must be compared with the cost of the production run for the system. Employing the minimum number of elements to perform a function also means a greater reliability (since there are fewer elements to break down) and a reduction in equipment size. Boolean algebra can be beneficial in providing the solution that encompasses all of these factors.

This chapter first introduces Venn diagrams and then uses these to show the derivation of the postulates and theorems connected with Boolean algebra.

Venn Diagrams

Given a problem of finding all the people in a group who are single *and* aged between 18 and 30, we can analyze this task further by drawing a diagram called a Venn Diagram in order to show all the categories of people in the group, as shown in figure 3-1.

All people are represented in the diagrams within the rectangle. Single people are represented by circle A, people aged 18 to 30 by circle B, and all people who are neither 18 to 30 nor single by area D. Thus all those people who are both single *and* 18 to 30 are represented by area C, and only these meet the specifications. This area is therefore the intersection of A and B, and may be written in algebraic form as $A \cap B$ (where $\cap$ is a symbol representing intersection) or as A AND B. For logic-design purposes the latter form is normally written $A \cdot B$, instead of A AND B. For ease of writing this in turn is sometimes reduced by not using the dot operator, and the expression then becomes AB.

The solution to the problem may now be written in algebraic form:

$$A \cdot B = C$$

Another definition pertaining to the Venn Diagram is that everything within the boundary of the rectangle is defined as the *universal set* U.

Now consider the problem of finding all people in the group who are single *or* aged 18 to 30. Figure 3-2 illustrates a Venn Diagram for this problem.

The shaded area C now relates to all the people who are single (A) *or* are aged 18 to 30 (B). This is the union of A and B and may be written as $A \cup B$ where $\cup$ represents *union*. Once again, for logic-design purposes this union symbol may be replaced by another, the OR connective. The area now becomes A OR B and is written as $A + B$.

The solution to the problem may be written in algebraic form as:

$$A + B = C$$

Therefore, in defining a Venn Diagram, the first two postulates of Boole's algebra have been proved. The others will be proven later in this chapter, in a similar but shortened manner, by first using Venn and then the basic logic gates to show the relationship between the algebra and the actual logic.

The two-variable Venn Diagram (A and B) shown in figures 3-1 and 3-2, can be extended to a three-variable diagram, as shown in figure 3-3. The different areas on the diagram have been labeled using the AND connective previously defined.

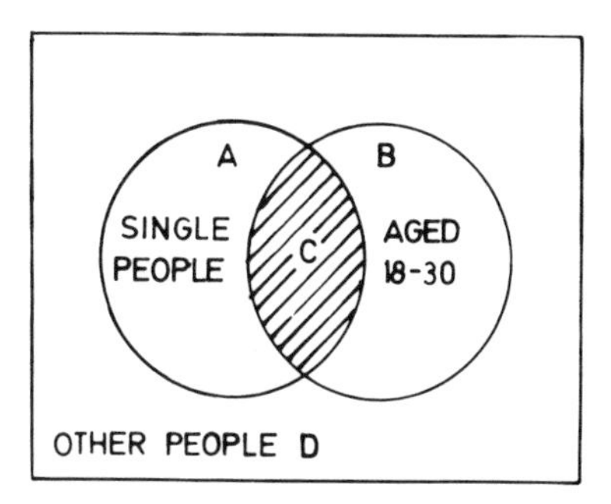

Figure 3–1. Venn Diagram of All People Who Are Single and Aged 18 to 30

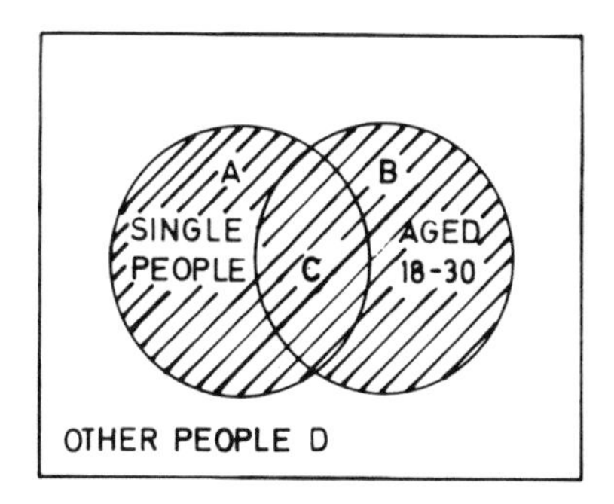

Figure 3–2. Venn Diagram of All People Who Are Single or Aged 18 to 30

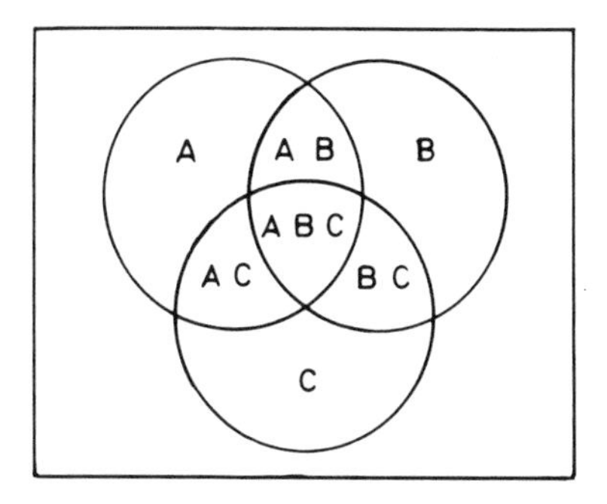

Figure 3–3. A Three-Variable Venn Diagram

Boolean Algebra Postulates

Each postulate is proved first by Venn and then by logic gates. The shaded area on the Venn Diagram represents the function above the Venn Diagram.

Postulate 1: A + B = C.

This is the symbolic notation for the statement: "If A or B are members of the universal set, then A + B is also a member of the Universal Set U.".

Proof 1:

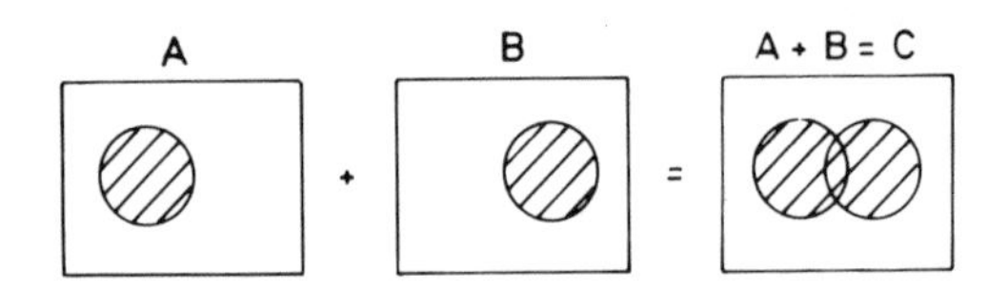

Proof 2: The OR Gate.

A	B	A + B = C
0	0	0
0	1	1
1	0	1
1	1	1

Postulate 2: A · B = C.

This is the symbolic notation for the statement: "If A and B are members of the universal set, then A · B is also the member of the universal set U."

Proof 1:

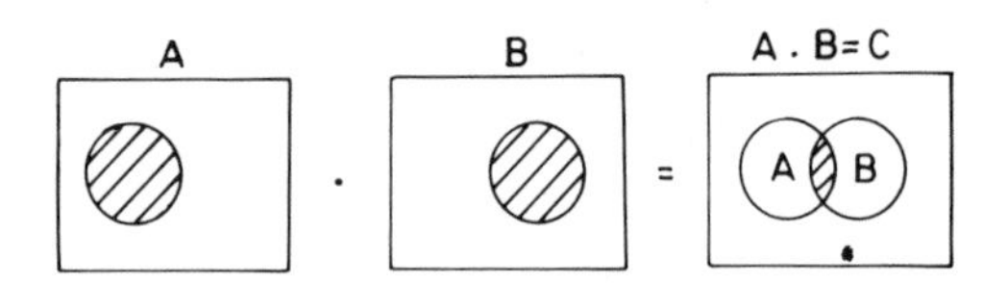

Proof 2: The AND Gate.

A	B	A · B = C
0	0	0
0	1	0
1	0	0
1	1	1

Postulate 3: $A + 0 = A$

There is an element, symbolized by 0 and known as the empty set, so that for any element A in the universal set U, $A + 0 = A$.

Proof 1:

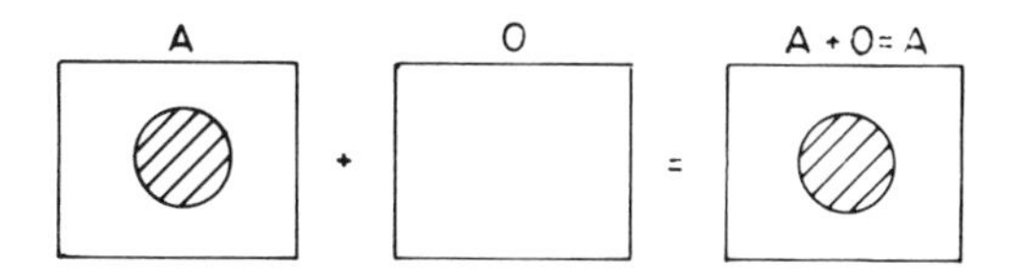

Proof 2:

A	0	A + 0 = A
0	0	0
1	0	1

Postulate 4: $A \cdot 1 = A$.

The universal set may be symbolized by 1 and is the set such that for any element A in the universal set, $A \cdot 1 = A$.

Proof 1:

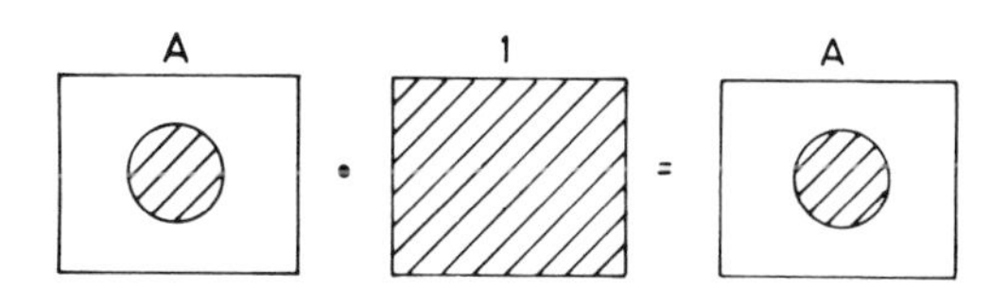

Proof 2:

A	B	A · 1 = A
0	1	0
1	1	1

Postulate 5: $A + B = B + A$

This is the symbolic notation for the statement "A + B is identical to B + A." This is known as the *first commutative law.*

Proof 1:

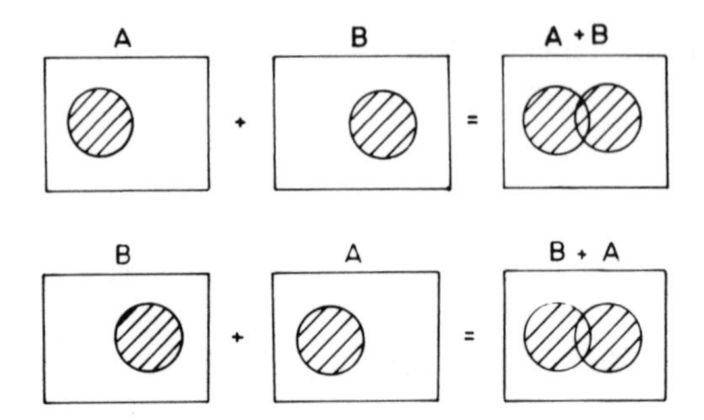

Proof 2:

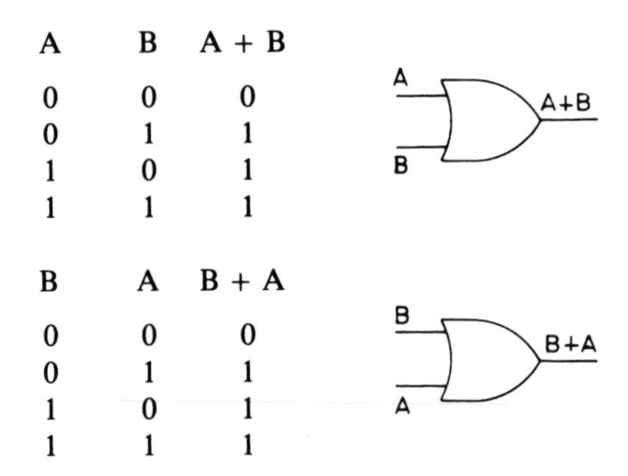

A	B	A + B
0	0	0
0	1	1
1	0	1
1	1	1

B	A	B + A
0	0	0
0	1	1
1	0	1
1	1	1

Postulate 6: $A \cdot B = B \cdot A$

This is the symbolic notation for the statement "A and B is identical to B and A." This is known as the *second commutative law.*

Proof 1:

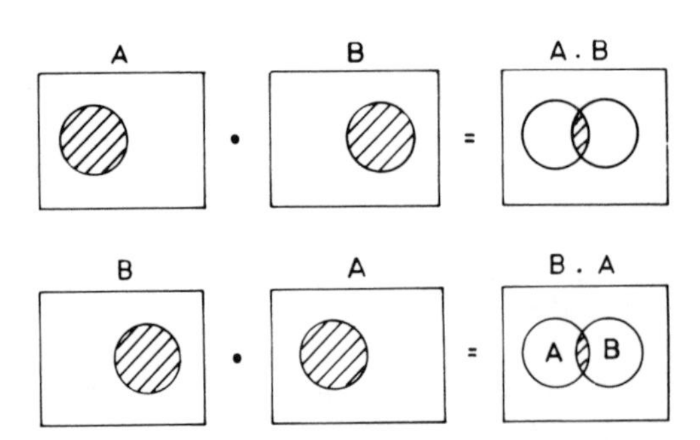

Proof 2:

A	B	A · B
0	0	0
0	1	0
1	0	0
1	1	1

B	A	B · A
0	0	0
0	1	0
1	0	0
1	1	1

Postulate 7: $A(B + C) = (A \cdot B) + (A \cdot C)$

This is known as the *first distributive law.*

Proof 1:

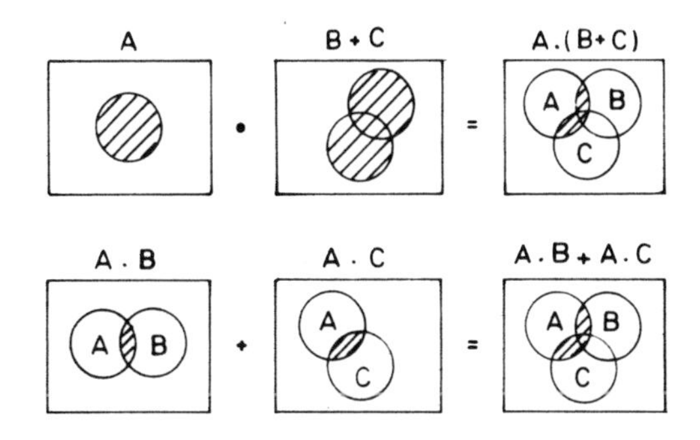

Proof 2:

A	B	C	B + C	A · (B + C)
0	0	0	0	0
0	0	1	1	0
0	1	0	1	0
0	1	1	1	0
1	0	0	0	0
1	0	1	1	1
1	1	0	1	1
1	1	1	1	1

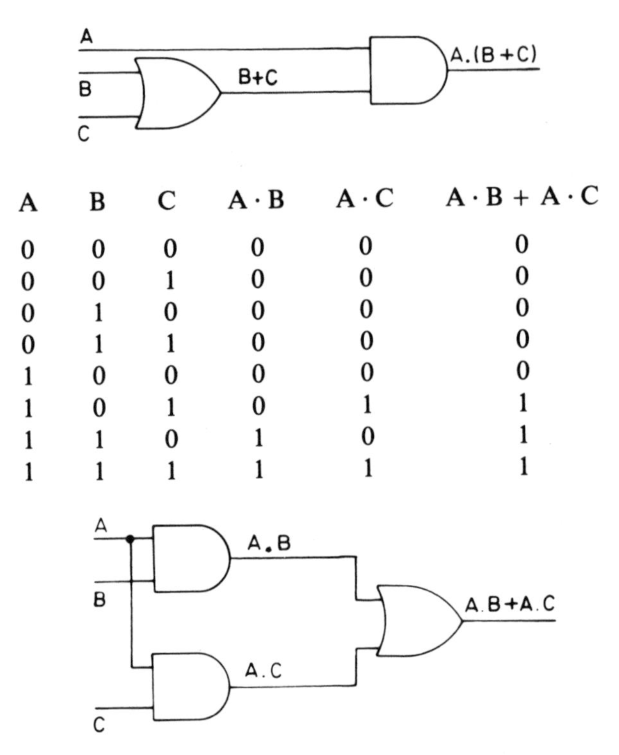

A	B	C	A · B	A · C	A · B + A · C
0	0	0	0	0	0
0	0	1	0	0	0
0	1	0	0	0	0
0	1	1	0	0	0
1	0	0	0	0	0
1	0	1	0	1	1
1	1	0	1	0	1
1	1	1	1	1	1

One should observe the second proof using the truth tables and logic gates, because using this postulate it is possible to rearrange the Boolean expression defining the logic function so that only two of the three gates are employed.

Postulate 8: $A + BC = (A + B)(A + C)$

This is known as the *second distributive law.*

Proof 1:

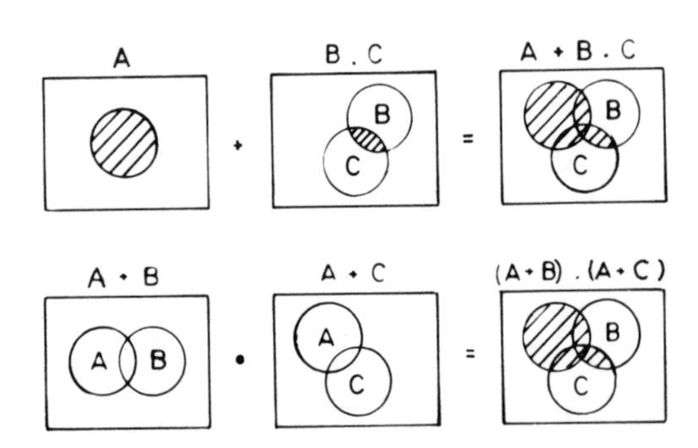

Proof 2:

A	B	C	B · C	A + B · C
0	0	0	0	0
0	0	1	0	0
0	1	0	0	0
0	1	1	1	1
1	0	0	0	1
1	0	1	0	1
1	1	0	0	1
1	1	1	1	1

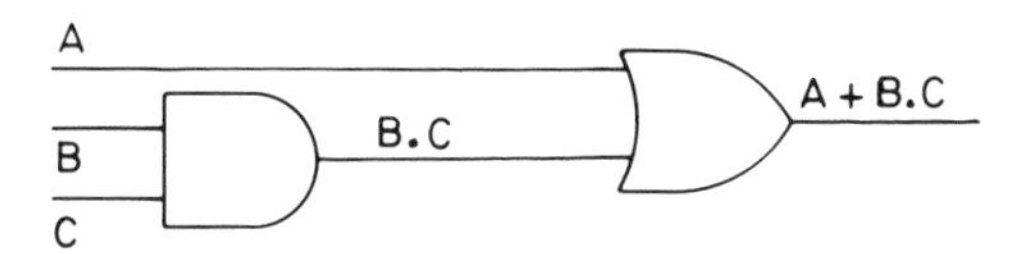

A	B	C	A + B	A + C	(A + B)(A + C)
0	0	0	0	0	0
0	0	1	0	1	0
0	1	0	1	0	0
0	1	1	1	1	1
1	0	0	1	1	1
1	0	1	1	1	1
1	1	0	1	1	1
1	1	1	1	1	1

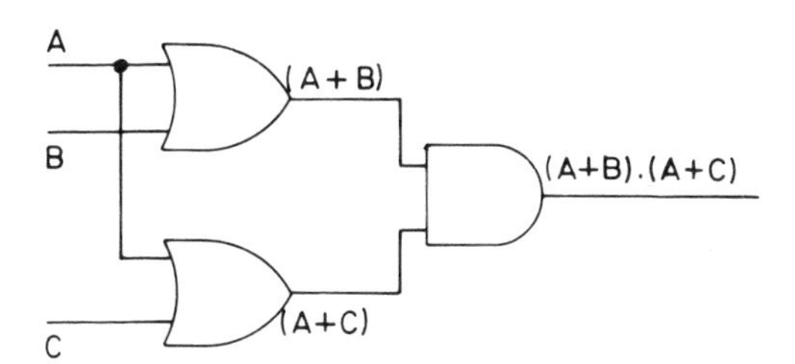

Once again it may be seen that by using the Second Distributive Law the number of logic gates required to perform the function may be reduced.

Postulate 9: $A + \bar{A} = 1$

This is symbolic notation for the statement "A or all elements that are not A, (that is, $\bar{A}$) includes all elements of the universal set." It is known as the *first law of complementation.*

Proof 1:

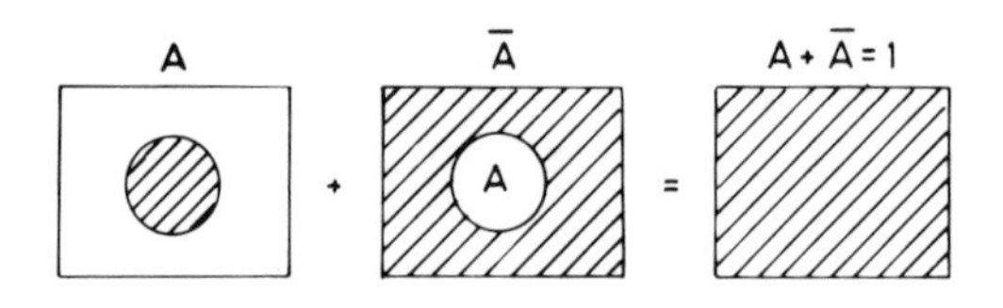

Proof 2:

A	$\bar{A}$	$A + \bar{A} = 1$
0	1	1
1	0	1

Postulate 10: $A \cdot \bar{A} = 0$

This is the symbolic notation for the statement: "The element A and all elements that are not A (that is, $\bar{A}$) includes only the zero element." It is known as the *second law of complementation.*

Proof 1:

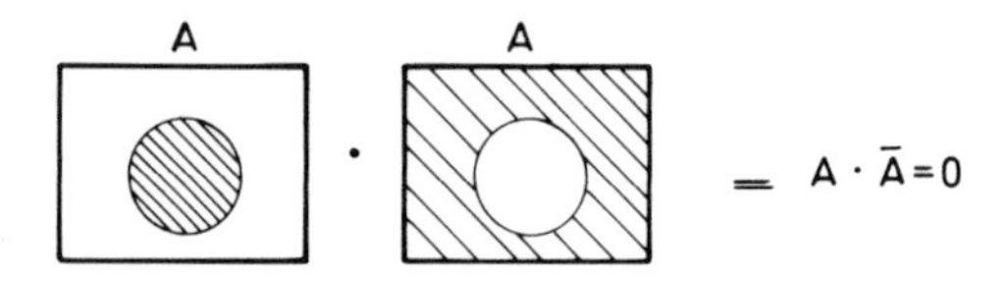

The Theorems of Boolean Algebra

In addition to the ten postulates of Boolean algebra that have been introduced, there are also twelve theorems that must be derived before the algebra can be used as an effective tool in the design of logic circuitry.

Three proofs are given for each theorem, the first using Venn Diagrams, the second employing the postulates, and the third with the aid of logic gates to show the practical usage of the theorem.

Theorem 1: $A + A = A$

Proof 1:

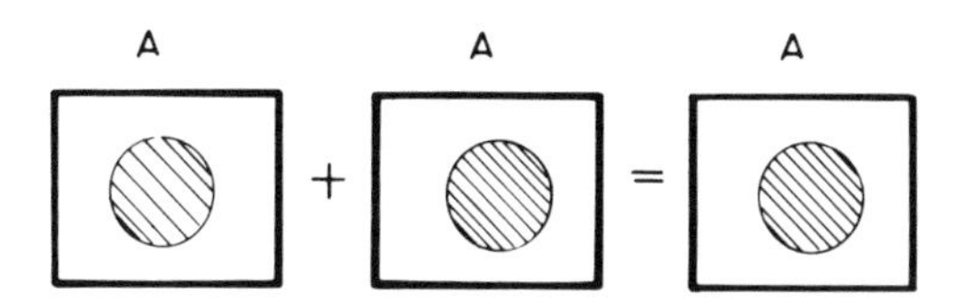

Proof 2:

$$A + A = (A + A) \cdot 1 \qquad \text{(by postulate 4)}$$
$$= (A + A) \cdot (A + \bar{A}) \qquad \text{(by postulate 9)}$$
$$= A + (A \cdot \bar{A}) \qquad \text{(by postulate 8)}$$
$$= A + 0 \qquad \text{(by postulate 10)}$$
$$= A \qquad \text{(by postulate 3)}$$

Proof 3:

A	A	A + A = A
0	0	0
1	1	1

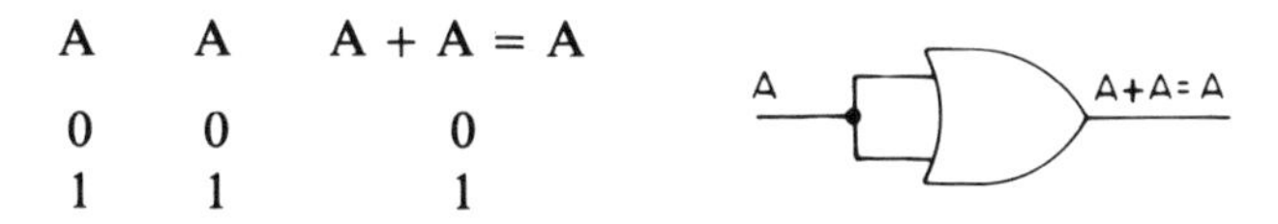

Theorem 2: $A \cdot A = A$

Proof 1:

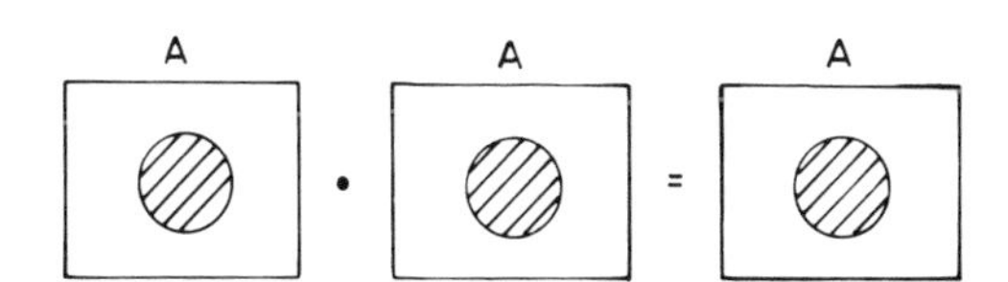

Proof 2:

$$A \cdot A = AA + 0 \qquad \text{(by postulate 3)}$$
$$= AA + A\bar{A} \qquad \text{(by postulate 10)}$$
$$= A(A + \bar{A}) \qquad \text{(by postulate 7)}$$
$$= A \cdot 1 \qquad \text{(by postulate 9)}$$
$$= A \qquad \text{(by postulate 4)}$$

Proof 3:

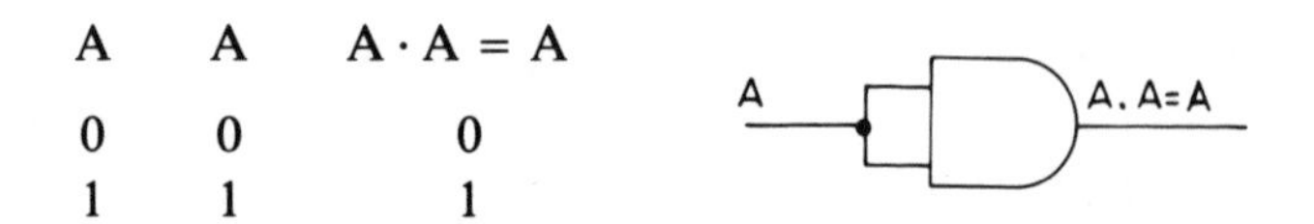

A	A	$A \cdot A = A$
0	0	0
1	1	1

Theorem 3: $A + 1 = 1$

Proof 1:

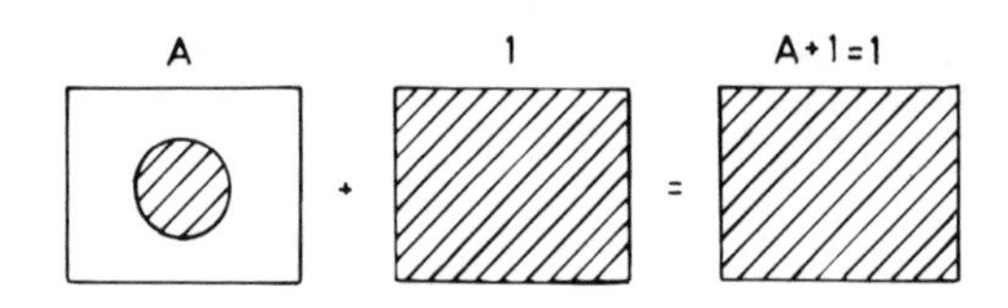

Proof 2:

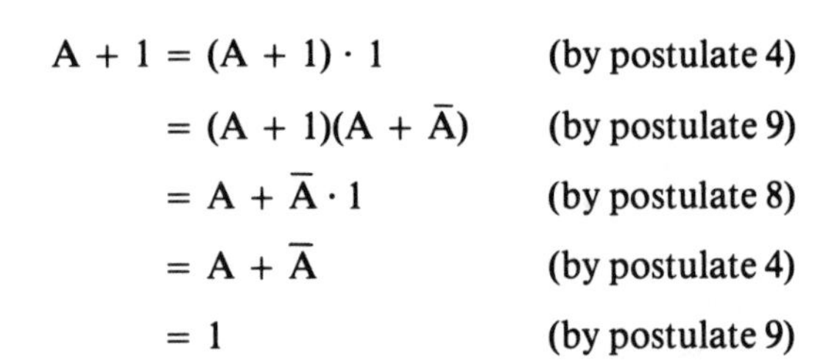

$$\begin{aligned} A + 1 &= (A + 1) \cdot 1 && \text{(by postulate 4)} \\ &= (A + 1)(A + \bar{A}) && \text{(by postulate 9)} \\ &= A + \bar{A} \cdot 1 && \text{(by postulate 8)} \\ &= A + \bar{A} && \text{(by postulate 4)} \\ &= 1 && \text{(by postulate 9)} \end{aligned}$$

Proof 3:

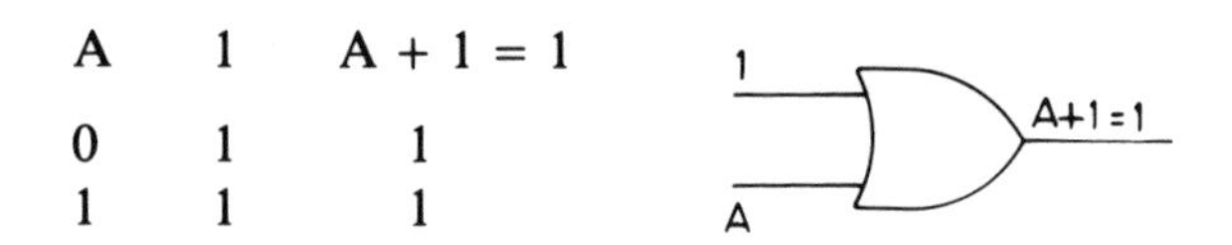

A	1	$A + 1 = 1$
0	1	1
1	1	1

Theorem 4: $A \cdot 0 = 0$

Proof 1:

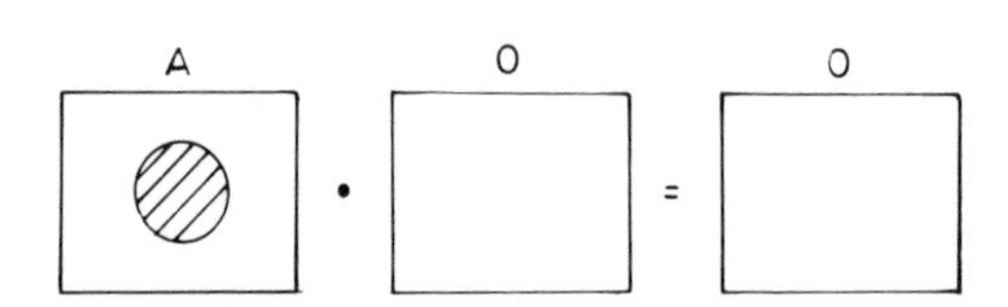

Proof 2:

$$\begin{aligned} A \cdot 0 &= A(A \cdot \overline{A}) && \text{(by postulate 10)} \\ &= (A + 0)(A \cdot \overline{A}) && \text{(by postulate 3)} \\ &= (A + 0)(A + 0)\overline{A} && \text{(by postulate 3)} \\ &= (A + 0)\overline{A} && \text{(by theorem 2)} \\ &= A\overline{A} && \text{(by postulate 3)} \\ &= 0 && \text{(by postulate 10)} \end{aligned}$$

Proof 3:

A	0	$A \cdot 0 = 0$
0	0	0
1	0	0

Theorem 5: $A + AB = A$

Proof 1:

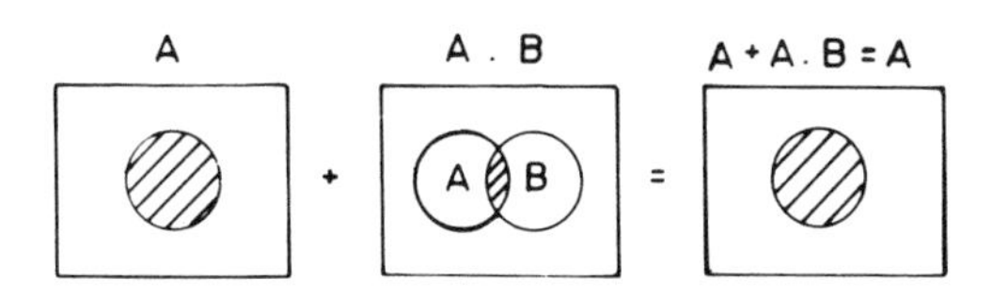

Proof 2:

$$\begin{aligned} A + AB &= A(1 + B) && \text{(by postulate 7)} \\ &= A \cdot 1 && \text{(by theorem 3)} \\ &= A && \text{(by postulate 4)} \end{aligned}$$

Proof 3:

A	B	$A \cdot B$	$A + A \cdot B = A$
0	0	0	0
0	1	0	0
1	0	0	1
1	1	1	1

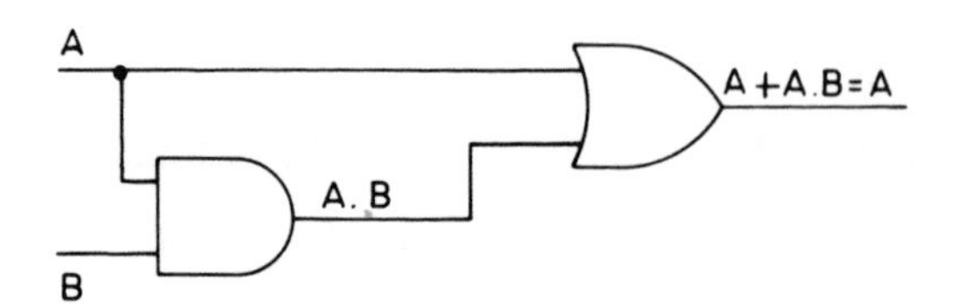

Thus the two gates are redundant, and only A is required to perform the function.

Theorem 6: $A(A + B) = A$

Proof 1:

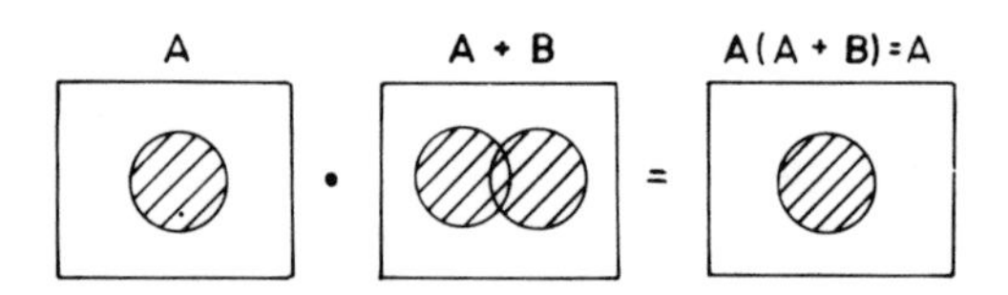

The area in the Venn Diagram covered by both A AND (A OR B) is only that of A, hence $A \cdot (A + B) = A$.

Proof 2:

$$\begin{aligned} A(A + B) &= AA + AB && \text{(by postulate 7)} \\ &= A + AB && \text{(by theorem 2)} \\ &= A(1 + B) && \text{(by postulate 7)} \\ &= A \cdot 1 && \text{(by theorem 3)} \\ &= A \end{aligned}$$

Proof 3:

A	B	A + B	A · (A + B) = A
0	0	0	0
0	1	1	0
1	0	1	1
1	1	1	1

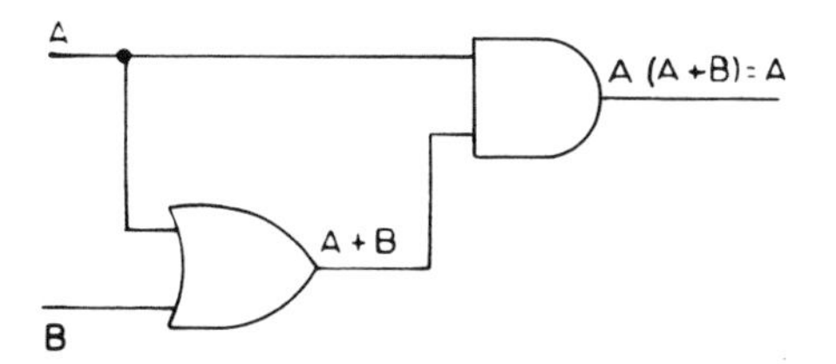

Therefore, once again the two gates are redundant, only A is required to perform the function.

Theorem 7: $\overline{(A + B)} = \bar{A} \cdot \bar{B}$ (DeMorgan's Theorem)

The simple rule for this theorem is "split the line and change the sign." Hence the expression $\overline{A + B}$ becomes $\bar{A} + \bar{B}$ when the line is split and then $\bar{A} \cdot \bar{B}$ when the sign is changed. It is always necessary to carry out both actions.

The reverse action is also possible; the rule for this is: "make the line and change the sign." Thus $\bar{A} \cdot \bar{B}$ becomes $\overline{A \cdot B}$ when the line is "made" and then $\overline{A + B}$ when the sign is changed. Once again, of course, both actions must be carried out.

Proof 1:

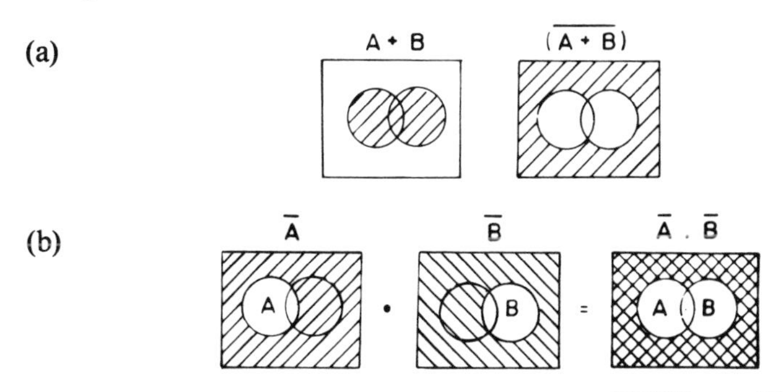

In (a) and (b) the same area is covered by $\overline{A + B}$ and $\bar{A} \cdot \bar{B}$; therefore, $\overline{A + B} = \bar{A} \cdot \bar{B}$.

Proof 2:

From the two Venn diagrams provided for $\bar{A} \cdot \bar{B}$ and A + B, it can be seen that one is the complement of the other. Consequently, if the two functions are ORed together, the result will be 1. Therefore, if $(A + B) + \bar{A} \cdot \bar{B} = 1$, then DeMorgans must be true.

This expression has been proven below by using the Theorems and Postulates derived so far and from this DeMorgan's Theorem has been proved.

$$A + \overline{A}\overline{B} + B = 1 \quad \text{(by postulate 8)}$$

$$(A + \overline{A})(A + \overline{B}) + B = 1 \quad \text{(by postulate 9)}$$

$$1 \cdot (A + \overline{B}) + B = 1 \quad \text{(by postulate 4)}$$

$$A + (\overline{B} + B) = 1$$

$$A + 1 = 1 \quad \text{(by postulate 9)}$$

$$1 = 1 \quad \text{(by theorem 3)}$$

Proof 3:

(a)

A	B	$\overline{A}$	$\overline{B}$	$\overline{A} \cdot \overline{B}$
0	0	1	1	1
0	1	1	0	0
1	0	0	1	0
1	1	0	0	0

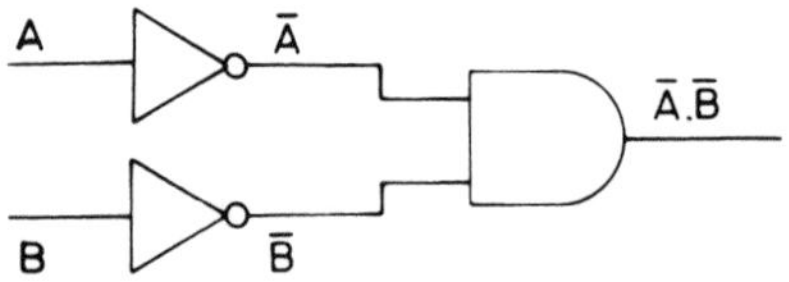

(b)

A	B	A + B	$\overline{A + B}$
0	0	0	1
0	1	1	0
1	0	1	0
1	1	1	0

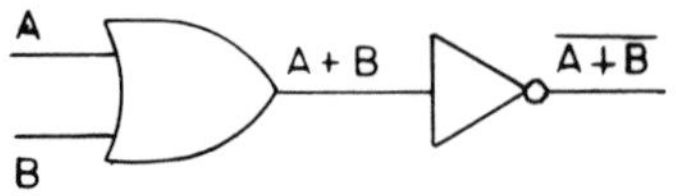

The two functions are equal, as illustrated by the truth tables. Also, the two logic circuits perform the same function but employ a different number and different types of gates.

Theorem 8: $(\overline{A \cdot B}) = \overline{A} + \overline{B}$ (DeMorgan's Theorem)

This theorem is another form of theorem 7, and the same rule of splitting the line and changing the sign may be applied. The proof is analogous to that of theorem 7, therefore only two proofs are given.

Proof 1:

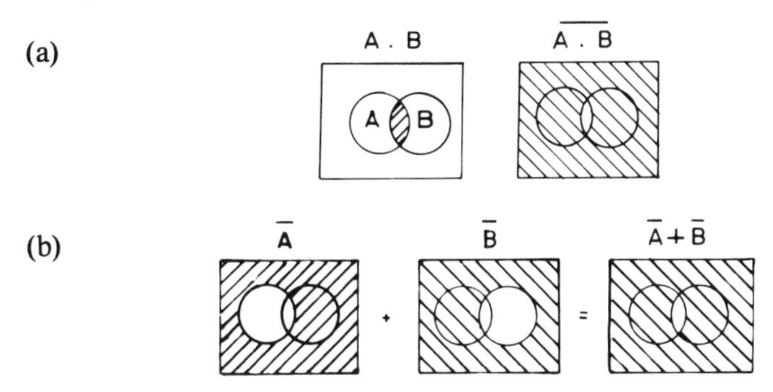

The areas covered by $\overline{A \cdot B}$ and $\overline{A} + \overline{B}$ are then the same, and therefore the expressions must be equal.

Theorem 9: $(\overline{\overline{A}}) = A$

Proof 1:

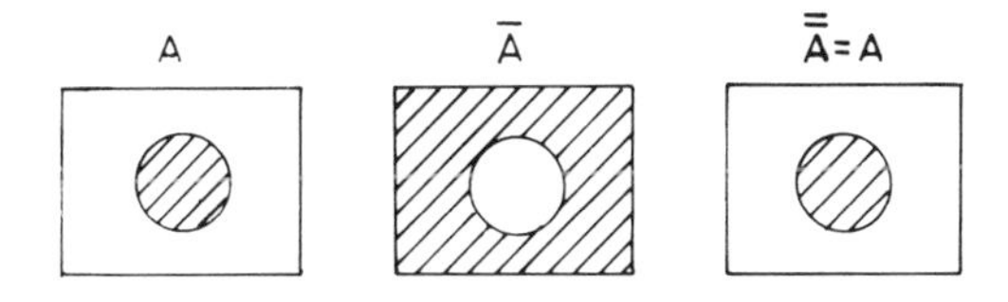

Proof 2:

A	$\overline{A}$	$\overline{\overline{A}} = A$
0	1	0
1	0	1

Theorem 10: $(A \cdot B) \cdot C = A \cdot (B \cdot C)$

This is the *first associative law.*

Proof 1:

(a) A . B · C = (A.B).C

(b) A · B . C = A .(B.C)

The same area is covered in (a) and (b) by both (A · B) · C and A · (B · C), and therefore these expressions are equal.

Proof 2:

(a)

A	B	C	(A · B)	(A · B) · C
0	0	0	0	0
0	0	1	0	0
0	1	0	0	0
0	1	1	0	0
1	0	0	0	0
1	0	1	0	0
1	1	0	1	0
1	1	1	1	1

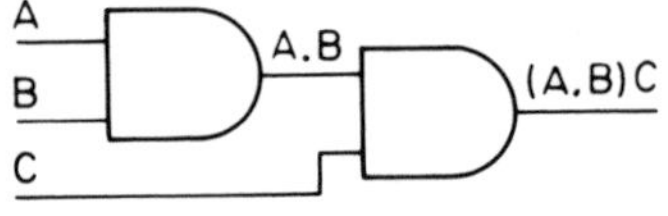

(b)

A	B	C	(B · C)	A(B · C)
0	0	0	0	0
0	0	1	0	0
0	1	0	0	0
0	1	1	1	0
1	0	0	0	0
1	0	1	0	0
1	1	0	0	0
1	1	1	1	1

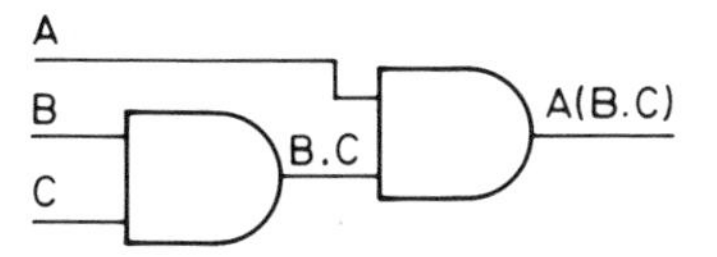

Theorem 11: (A + B) + C = A + (B + C)

This theorem is the *second associative law.*

Proof 1:

(a) A +B + C = (A + B)+C

(b) A + B+C = A+(B +C)

The same area is covered in (a) and (b) by both (A + B) + C and A + (B + C), and therefore these expressions are equal.

Proof 2:

(a)

A	B	C	(A + B)	(A + B) + C
0	0	0	0	0
0	0	1	0	1
0	1	0	1	1
0	1	1	1	1
1	0	0	1	1
1	0	1	1	1
1	1	0	1	1
1	1	1	1	1

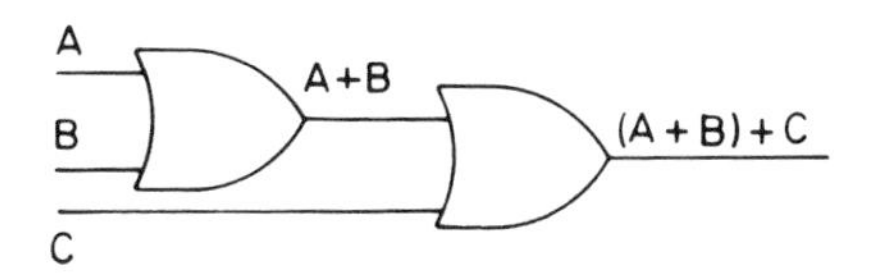

(b)

A	B	C	B + C	A + (B + C)
0	0	0	0	0
0	0	1	1	1
0	1	0	1	1
0	1	1	1	1
1	0	0	0	1
1	0	1	1	1
1	1	0	1	1
1	1	1	1	1

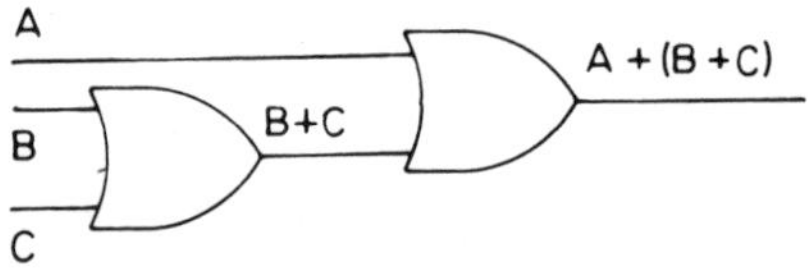

The truth tables are the same in each case, hence the expressions are equal.

Theorem 12: $A + \overline{A}B = A + B$

Proof 1:

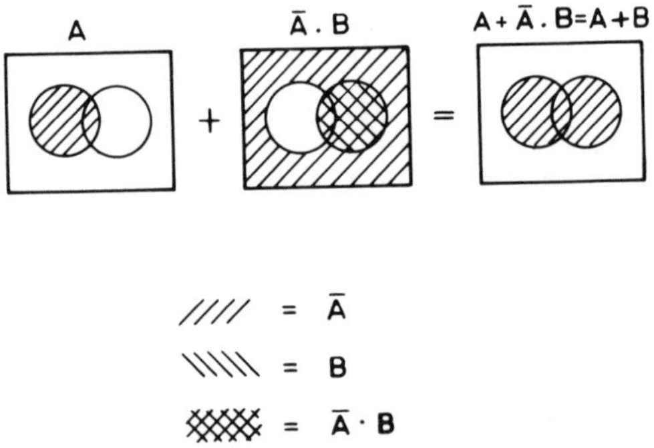

Proof 2:

$$A + \overline{A} \cdot B = A + B$$

$$(A + \overline{A}) \cdot (A + B) = A + B \qquad \text{(by postulate 8)}$$

$$1 \cdot (A + B) = A + B \qquad \text{(by postulate 9)}$$

$$A + B = A + B \qquad \text{(by postulate 4)}$$

Proof 3:

A	B	$\bar{A}$	$\bar{A}B$	$A + \bar{A}B$	$= A + B$
0	0	1	0	0	0
0	1	1	1	1	1
1	0	0	0	1	1
1	1	0	0	1	1

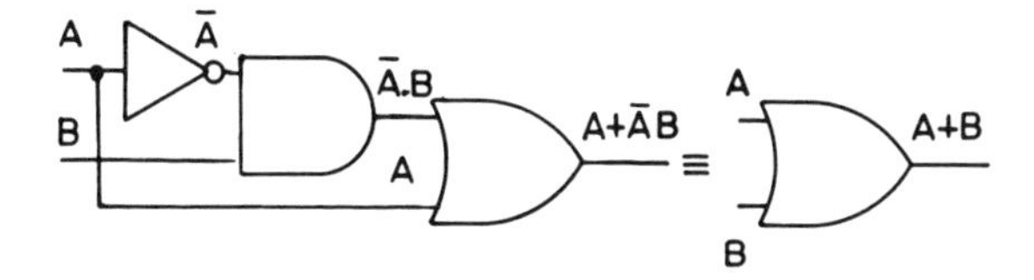

To perform the function in this instance the logic circuit may be reduced from three gates to just one OR gate.

Summary of Postulates and Theorems

Postulates

3	$A + 0 = A$	
4	$A \cdot 1 = A$	
5	$A + B = B + A$	First commutative law
6	$A \cdot B = B \cdot A$	Second commutative law
7	$A(B + C) = (A \cdot B) + (A \cdot C)$	First distributive law
8	$A + BC = (A + B)(A + C)$	Second distributive law
9	$A + \bar{A} = 1$	First law of complementation
10	$A \cdot \bar{A} = 0$	Second law of complementation

Theorems

1 $A + A = A$

2 $A \cdot A = A$

3 $A + 1 = 1$

4 $A \cdot 0 = 0$

5	$A + AB = A$	
6	$A(A + B) = A$	
7	$\overline{(A + B)} = \overline{A}\overline{B}$	DeMorgan's Theorem
8	$\overline{(A \cdot B)} = \overline{A} + \overline{B}$	DeMorgan's Theorem
9	$(\overline{\overline{A}}) = A$	
10	$(A \cdot B)C = A(B \cdot C)$	First associative law
11	$(A + B) + C = A + (B + C)$	Second associative law
12	$A + \overline{A}B = A + B$	

Simplifying Boolean Expressions Using the Postulates and Theorems

Boolean algebra has a very important role in the design of logic circuits. A Boolean expression may be derived from the truth table for a particular function, as will be shown later. Once the expression has been derived, an attempt must be made to simplify it. If it can be simplified, then a more economical logic network will nearly always result. The technique of simplifying Boolean expressions is termed *minimization*.

Example 3.1

Simplify the following Boolean function:

$$f = A + \overline{A}B + AB$$

With many Boolean expressions there may be more than one method of reduction, but the same result will be obtained. Occasionally an equally correct, although different, result may be obtained. One method of simplification may be quicker than another, and learning to choose the quickest can be achieved only by practice. For these reasons this example will be simplified using two different methods.

Method 1:

$$\begin{aligned} f &= A + \overline{A}B + AB \\ &= A + B(\overline{A} + A) \quad &\text{(by postulate 7)} \\ &= A + B \quad &\text{(by postulate 9)} \end{aligned}$$

Method 2:

$$
\begin{aligned}
f &= A + \bar{A}B + AB \\
&= (A + \bar{A}B) + AB \\
&= A + B + AB && \text{(by theorem 12)} \\
&= A(B + 1) + B && \text{(by postulate 7)} \\
&= A \cdot 1 + B && \text{(by theorem 3)} \\
&= A + B && \text{(by postulate 4)}
\end{aligned}
$$

Although both methods give the same result, in this case method 1 was the shorter.

Example 3.2

Simplify:

$$f = A\bar{C}D + \bar{A}CD + ACD$$

Method 1:

$$
\begin{aligned}
f = A\bar{C}D + \bar{A}CD + ACD &= A\bar{C}D + (\bar{A} + A)CD && \text{(by postulate 7)} \\
&= A\bar{C}D + CD && \text{(by postulate 9)} \\
&= D(\bar{C}A + C) && \text{(by postulate 7)} \\
&= D(A + C) && \text{(by theorem 12)} \\
&= AD + AC
\end{aligned}
$$

Method 2:

$$
\begin{aligned}
f = A\bar{C}D + \bar{A}CD + ACD &= AD(\bar{C} + C) + \bar{A}CD && \text{(by postulate 7)} \\
&= AD + \bar{A}CD && \text{(by postulate 9)} \\
&= D(A + \bar{A}C) && \text{(by postulate 7)} \\
&= D(A + C) && \text{(by theorem 12)} \\
&= AD + AC
\end{aligned}
$$

Example 3.3

$$
\begin{aligned}
f &= \overline{A}BC + \overline{A}\overline{B}C + \overline{A}B\overline{C} + ABC + AB\overline{C} \\
&= BC(A + \overline{A}) + B\overline{C}(A + \overline{A}) + \overline{A}\cdot\overline{B}\cdot C \\
&= BC + B\overline{C} + \overline{A}\cdot\overline{B}\cdot C && \text{(by postulate 9)} \\
&= B(C + \overline{C}) + \overline{A}\cdot\overline{B}\cdot C \\
&= \overline{A}\cdot\overline{B}\cdot C + B \\
&= \overline{A}\cdot C + B && \text{(by theorem 12)}
\end{aligned}
$$

Example 3.4

Boolean algebra can be used not only to simplify circuits but also to construct a circuit performing the same function as another but using different gates. This is demonstrated by this example.

Draw a circuit symbol equivalent to:

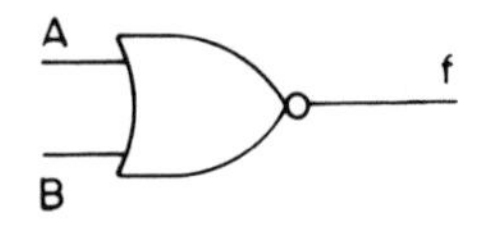

Solution: This symbol is that of the NOR gate, and therefore the function

$$
\begin{aligned}
f &= \overline{A + B} \\
&= \overline{A}\cdot\overline{B} && \text{by DeMorgan's Theorem}
\end{aligned}
$$

The circuit for $\overline{A}\cdot\overline{B}$ can be drawn as:

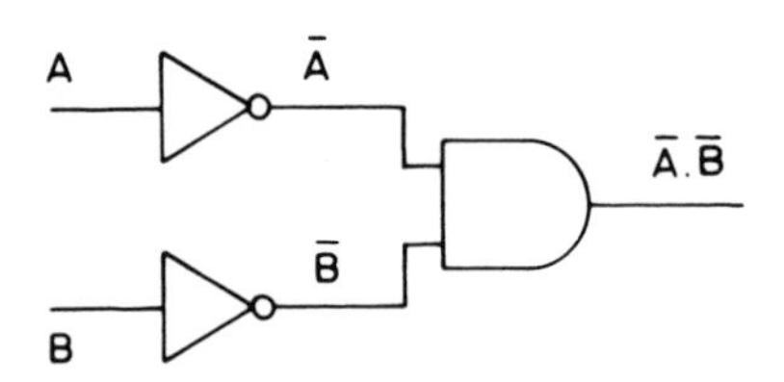

which can alternatively be drawn as:

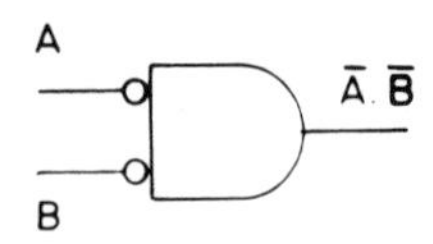

Example 3.5

Simplify the following circuit to fewer gates.

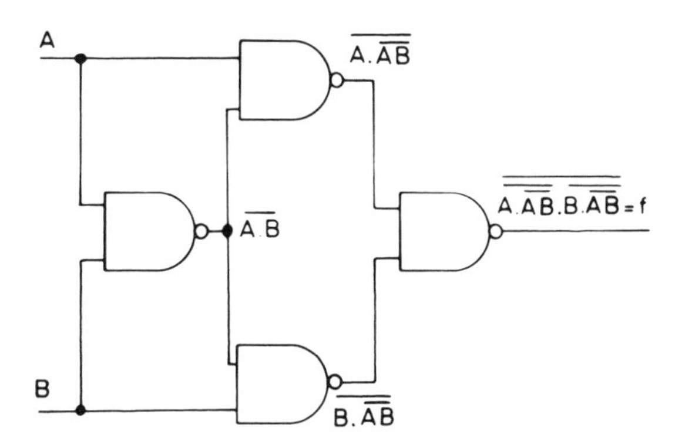

Solution: The Boolean expression for the functions at each stage have been entered onto the diagram; hence the overall expression for the circuit output f is derived.

$$f = \overline{\overline{A\,\overline{AB}}\cdot\overline{B\,\overline{AB}}}$$

$$= \overline{\overline{A\cdot\overline{AB}}} + \overline{\overline{B\cdot\overline{AB}}} \qquad \text{(by DeMorgan's Theorem)}$$

$$= A\cdot\overline{AB} + B\cdot\overline{AB} \qquad \text{(by theorem 9)}$$

$$= \overline{A\cdot B}\cdot(A + B) \qquad \text{(by distributive law)}$$

$$(\bar{A} + \bar{B})\cdot(A + B) \qquad \text{(by DeMorgan's Theorem)}$$

$$= A\cdot\bar{A} + \bar{B}\cdot A + \bar{A}\cdot B + \bar{B}\cdot B \qquad \text{(by distributive law)}$$

$$= A\bar{A} + \bar{B}A + \bar{A}B + \bar{B}B$$

Therefore:

$$f = \bar{B}A + \bar{A}B \qquad \text{(by postulate 10)}$$

The logic circuit may now be drawn for this function, as follows:

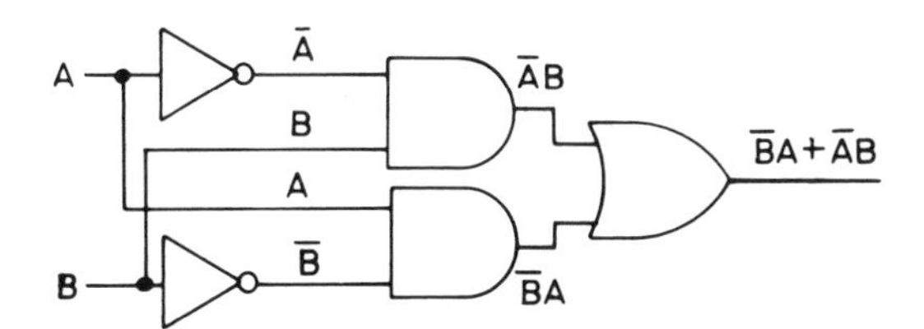

In this case, although the Boolean expression for the function is simpler than the original, the logic circuit is not. This means that in a few problems, such as the one just shown, the simplification of the expression does not lead to a simpler circuit. In fact, in this instance the first circuit would probably be the cheaper of the two.

However, if the truth table for the simplified function $f = \overline{B}A + \overline{A}B$ is drawn next, then a simpler result can be derived from it:

A	B	$\overline{A}$	$\overline{B}$	$\overline{B} \cdot A$	$\overline{A} \cdot B$	$\overline{B}A + \overline{A}B$
0	0	1	1	0	0	0
0	1	1	0	0	1	1
1	0	0	1	1	0	1
1	1	0	0	0	0	0

It can be seen that this is actually the truth table for an EXCLUSIVE OR gate and therefore the equivalent circuit is:

In algebraic form function f may be written:

$$f = A \oplus B$$

where the $\oplus$ sign represents the EXCLUSIVE OR function.

Exercise 3.1

Minimize the following expressions using the postulates and theorems.

1. $f = A \cdot \overline{B} \cdot \overline{C} + A \cdot B \cdot \overline{C} + A \cdot \overline{B} \cdot C + A \cdot B \cdot C$
2. $f = A \cdot B \cdot \overline{C} + \overline{A} \cdot \overline{B} \cdot \overline{C} + A \cdot \overline{B} \cdot \overline{C} + \overline{A} \cdot B \cdot \overline{C}$
3. $f = A \cdot B \cdot C + B \cdot C + A \cdot \overline{B} \cdot C$
4. $f = A \cdot \overline{B} + \overline{A} \cdot \overline{B} \cdot C + \overline{A} \cdot \overline{B} \cdot \overline{C}$

Exercise 3.2

Minimize the following expressions using the postulates and theorems.

1. $f = \overline{A} \cdot \overline{B} \cdot C \cdot \overline{D} + \overline{A} \cdot B \cdot C \cdot D + A \cdot B \cdot \overline{C} \cdot D + A \cdot B \cdot C \cdot D$
   $+ A \cdot \overline{B} \cdot \overline{C} \cdot D + A \cdot \overline{B} \cdot C \cdot \overline{D}$

2. $f = A \cdot B \cdot \overline{C} \cdot \overline{D} + A \cdot \overline{B} \cdot C \cdot D + A \cdot B \cdot \overline{C} \cdot D + A \cdot \overline{B} \cdot \overline{C} \cdot D$
   $+ A \cdot B \cdot C \cdot D$

3. $f = \overline{A} \cdot B \cdot C \cdot \overline{D} + A \cdot B \cdot C \cdot \overline{D} + \overline{A} \cdot B \cdot \overline{C} \cdot \overline{D} + \overline{A} \cdot B \cdot C \cdot D$
   $+ A \cdot B \cdot C \cdot D + A \cdot B \cdot \overline{C} \cdot \overline{D}$

Modification of Expressions to Perform the Function Using NAND Gates Only

The solutions obtained in the previous examples and exercises may all be realized using AND, NOT, and OR gates. Although the solutions may employ the minimum number of gates, they may not be the most cost effective. In general NAND gates are the cheapest, and a solution constructed from NAND gates may be best.

It should be pointed out, however, that the number of packages and the overall cost are the important factors in determining the most cost-effective circuit, rather than simply the type of gates utilized.

Expressions can be converted to NAND-gate form with the aid of DeMorgan's Theorem as shown in example 3.6.

Example 3.6

Convert the solution of example 3.2 (f = AD + AC) to NAND-gate form. (Note that this function has already been minimized.)

Solution: The solution at present is realized by the following circuit:

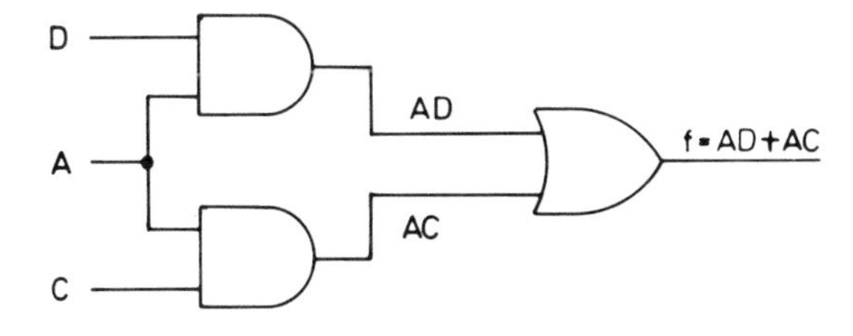

Conversion to NAND gates is as follows:

$f = AD + AC$

$\overline{f} = \overline{AD + AC}$ inverting both sides of the expression

$\overline{f} = \overline{AD} \cdot \overline{AC}$ by DeMorgan's Theorem

$f = \overline{\overline{AD} \cdot \overline{AC}}$ inverting both sides of the expression

Thus the function may be realized using NAND gates as follows:

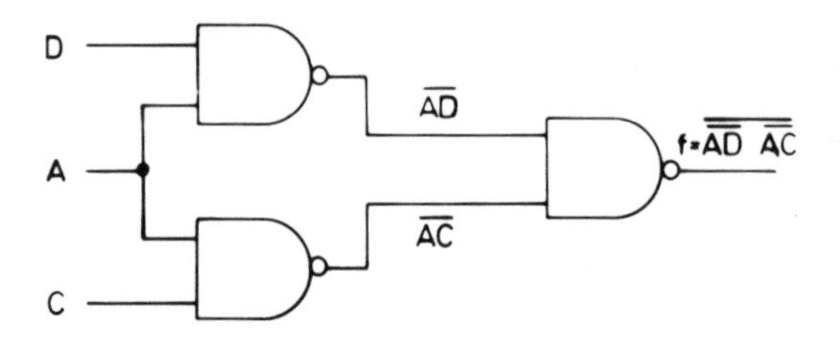

For this example the function may be realized by three NAND gates, which could all be part of one 7400 NAND-gate package. The original circuit would require two packages (one AND-gate package and one OR-gate package) and would consequently cost more.

Exercise 3.3

1. Perform the following function using one 7400 NAND-gate package only:

$$f = A \cdot B \cdot C + \overline{A} \cdot \overline{B} \cdot C + A \cdot \overline{B} \cdot C$$

2. Perform the following function using two NAND-gate packages only:

$$f = \overline{A} \cdot \overline{B} \cdot \overline{C} + A \cdot B \cdot \overline{C} + \overline{A} \cdot B \cdot \overline{C} + \overline{A} \cdot B \cdot C + A \cdot B \cdot C$$

Definition of Terms Used in Boolean Algebra

1. A *literal* refers to a variable or its complement, for example, A and A.
2. A *product term* is a group of literals ANDed together.
3. A *sum term* is a group of literals ORed together.
4. A *normal term* is a product or sum term in which a literal may appear more than once.
5. A *canonical term* of a function is a term containing exactly one occurrence of each literal of the function, for example:

$$f(A,B,C) = A \cdot B \cdot C + \overline{A} \cdot \overline{B} \cdot \overline{C}$$

6. A *sum of products* is product terms ORed together.
7. A *product of sums* is sum terms ANDed together.

The Canonical-Expansion Theorem

Expressions for functions may not only be minimized but may also be rearranged into other forms, including sum-of-products form or product-of-sums form. If the resulting expression contains only one occurrence of each literal, then it is also said to be a *canonical* form. An expression may be modified to sum-of-product canonical form or product-of-sum canonical form using the *canonical-expansion theorem.*

Consider a function f which is dependent on a selection of variables A to Z (A, B, C, . . . Z). This expression may be converted to sum-of-product canonical form by ANDing it with $(A + \overline{A}) \cdot (B + \overline{B}) \ldots (Z + \overline{Z})$, which is effectively the same as multiplying it by 1. The distributive axiom is then applied, as demonstrated by the following example:

Example 3.7

Obtain the sum-of-product canonical form of the following function:

$$f = A\overline{C} + \overline{B}C$$

Solution:

$$\begin{aligned} f &= A\overline{C} + \overline{B}C \\ &= (A\overline{C} + \overline{B}C) \cdot (A + \overline{A}) \cdot (B + \overline{B}) \cdot (C + \overline{C}) && \text{(by postulate 9)} \\ &= A\overline{C} \cdot (B + \overline{B}) + \overline{B}C \cdot (A + \overline{A}) \\ &= AB\overline{C} + A\overline{B}\overline{C} + A\overline{B}C + \overline{A}\overline{B}C && \text{(by postulate 7)} \end{aligned}$$

Note that as C or its complement appeared in both parts of the original expression, $(C + \overline{C})$ did not have to be ANDed with $A\overline{C}$ or $\overline{B}C$ to result in a sum-of-product canonical form.

To obtain the product-of-sum canonical form, the function $f(A, B, C, \ldots Z)$ must be ORed with $A\overline{A}$, $B\overline{B}$, $C\overline{C} \ldots Z\overline{Z}$, which is effectively the same as having 0 added. This is demonstrated by example 3.8.

Example 3.8

Obtain the product-of-sum canonical form of the following function:

$$f = (A + B) \cdot (B + C)$$

Solution:

$$f = (A + B) \cdot (B + C)$$
$$= (A + B) \cdot (B + C) + A\bar{A} + B\bar{B} + C\bar{C} \qquad \text{(by postulate 10)}$$

Let $X = A\bar{A} + B\bar{B} + C\bar{C}$; $Y = A + B$; and $Z = B + C$. Then:

$$f = YZ + X$$
$$= (Y + X) \cdot (Z + X) \qquad \text{(by postulate 8)}$$
$$= (A + B + A\bar{A} + B\bar{B} + C\bar{C}) \cdot (B + C + A\bar{A} + B\bar{B} + C\bar{C})$$
by resubstitution for X, Y, and Z
$$= (A + B + C\bar{C}) \cdot (B + C + A\bar{A}) \qquad \text{after removal of redundant terms}$$

Let $Y = A + B$ and $Z = B + C$ again. Then:

$$f = (X + C\bar{C}) \cdot (Z + A\bar{A})$$
$$= (X + C) \cdot (X + \bar{C}) \cdot (Z + A)(Z + \bar{A}) \qquad \text{(by postulate 8)}$$
$$= (A + B + C) \cdot (A + B + \bar{C})(A + B + C) \cdot (\bar{A} + B + C)$$ by resubstitution of Y and Z

(A + B + C) appears twice: hence it is a redundant pair and the expression becomes:

$$f = (A + B + C) \cdot (A + B + \bar{C}) \cdot (\bar{A} + B + C)$$

Thus it is now in its product-of-sum canonical form.

Logic-Design Synthesis

In order to design a circuit, one must first define its function and then construct a truth table for it. Once the truth table has been derived, a Boolean expression in either sum-of-product canonical form or product-of-sum canonical form is then obtained.

Derivation of Expression in Sum-of-Product Form from Truth Table

The rule for derivation of sum-of-product form is:

1. For each row of the table when the function is to be 1, synthesize the product term from the literals that will be 1 for only the values of the variables in that row.
2. All variables with value 0 are complemented; all those with value 1 are left unchanged.
3. The resulting literals are then ANDed together.
4. Finally these terms are then ORed to obtain the logical function.

This is demonstrated by example 3.9.

Example 3.9

Obtain the sum-of-product form of the function, given the following truth table:

Row	A	B	C	*f*
0	0	0	0	0
1	0	0	1	1
2	0	1	0	1
3	0	1	1	0
4	1	0	0	1
5	1	0	1	0
6	1	1	0	1
7	1	1	1	0

Solution: Look for terms to produce 1s, complement all variables with value 0, and leave all variables with 1 unchanged, as stated in the rule. The literals are then ANDed together.

The function is a 1 for:

Row 1 when A is 0, written $\bar{A}$ and B is 0, written $\bar{B}$ and C is 1, written C } $\bar{A}\bar{B}C$

Row 2 when A is 0, written $\bar{A}$ and B is 1, written B and C is 0, written $\bar{C}$ } $\bar{A}B\bar{C}$

Row 4 when A is 1, written A and B is 0, written $\bar{B}$ and C is 0, written $\bar{C}$ } $A\bar{B}\bar{C}$

Row 6 when A is 1, written A and B is 1, written B and C is 0, written $\bar{C}$ } $AB\bar{C}$

These products must now all be ORed together to give:

$$f = \overline{A}\,\overline{B}C + \overline{A}B\overline{C} + A\overline{B}\,\overline{C} + AB\overline{C}$$

the sum of product form for the function.

From this expression a logic circuit may be drawn directly or, alternatively, it may be minimized first.

Derivation of Product of Sum Form from Truth Table

The rule for the derivation of product-of-sum form is:

For each row of the table for which the function is a 0, synthesize a sum term that will be zero for only the values of the variables in that row. All the variables with value 1 are complemented while all the variables with value 0 are left unchanged.

The resulting literals are summed and the terms multiplied together.

This is demonstrated by example 3.10.

Example 3.10

Using the truth table given in example 3.9 obtain the product-of-sum form of the function.

Solution. Look for terms to produce 0s, complement all variables with value 1 and leave all variables with value 0 unchanged as stated in the rule. The literals are then ORed together.

The function is 0 for:

Row 0	when A is 0, written A or B is 0, written B or C is 0, written C	$A + B + C$
Row 3	when A is 0, written A or B is 1, written $\overline{B}$ or C is 1, written $\overline{C}$	$A + \overline{B} + \overline{C}$
Row 5	when A is 1, written $\overline{A}$ or B is 0, written B or C is 1, written $\overline{C}$	$\overline{A} + B + \overline{C}$
Row 7	when A is 1, written $\overline{A}$ or B is 1, written $\overline{B}$ or C is 1, written $\overline{C}$	$\overline{A} + \overline{B} + \overline{C}$

These sums must now be ANDed together to give:

$$f = (A + B + C) \cdot (A + \overline{B} + \overline{C}) \cdot (\overline{A} + B + \overline{C}) \cdot (\overline{A} + \overline{B} + C)$$

This is the product-of-sum form for the function.

Example 3.11

Design a logic circuit that allows a light to be independently controlled from three switches.

Solution: The circuit has three inputs. Call these A, B, and C. Let 0 = "switch up" and 1 = "switch down." Also let 0 = "light off" and 1 = "light on." Given the problem, a truth table can now be drawn as follows:

A	B	C	*f*
0	0	0	0
0	0	1	1
0	1	0	1
0	1	1	0
1	0	0	1
1	0	1	0
1	1	0	0
1	1	1	1

That is, for *one* switch down, the lamp will be on. For *two* switches down, the second cancels the effect of the first and therefore the lamp is off. For *three* switches down, the lamp will be on.

The sum-of-products form for the function can now be derived from the truth table, as previously demonstrated.

$$f = \overline{A}\overline{B}C + \overline{A}B\overline{C} + A\overline{B}\overline{C} + ABC$$

Unfortunately, this expression cannot be reduced any further, but it may be written alternatively as:

$$f = \overline{A}(\overline{B}C + B\overline{C}) + A(\overline{B}\overline{C} + BC)$$

The logic circuit consists of six AND gates (as there are six dots in the expression) and three OR gates (as three + signs). Three inverters will also be required to generate $\overline{A}$, $\overline{B}$, and $\overline{C}$. Therefore, a circuit that will perform the function is:

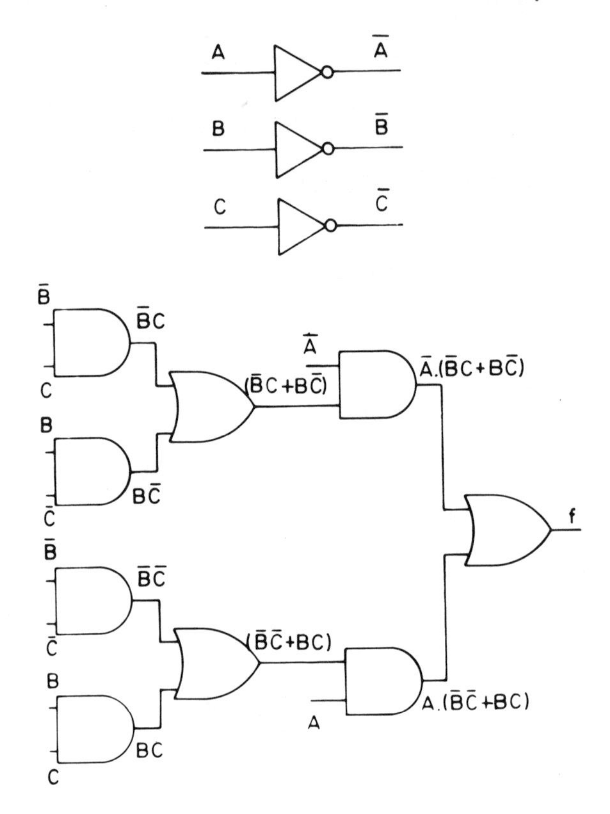

Minterms, Maxterms, and Don't Care's

A product term that contains all the variables of a function, either in their true or their negated form, is called a *minterm* (m). The following product terms are all examples of minterms: $\bar{A}\bar{B}\bar{C}$, $\bar{A}\bar{B}C$, and $\bar{A}B\bar{C}$. These terms may be related to a truth table, as demonstrated previously. Each minterm is then given a number that equates to the decimal equivalent of the 1s and 0s, as demonstrated by the following table:

A	B	C	*Minterm*
0	0	0	m_0
0	0	1	m_1
0	1	0	m_2
0	1	1	m_3
1	0	0	m_4
1	0	1	m_5
1	1	0	m_6
1	1	1	m_7

Therefore, an alternative method of writing a function f, such as $f = \bar{A}B\bar{C} + A\bar{B}\bar{C} + A\bar{B}C$, is as a sum of minterms, that is, for this example:

$$f = m_2 + m_4 + m_5$$

It is also possible to write a function in *maxterm* (M) form. A maxterm is a sum term that contains all the variables of a function, either in their true or their negated form. For example, the function $f = (A + B + C) \cdot (A + B + \bar{C}) \cdot (A + \bar{B} + C)$ may be expressed using maxterms as $f = M_7, M_6, M_5$. Note that maxterm 7 (M_7) actually equates to minterm 0 (m_0), M_6 to m_1, M_5 to m_2, and so on. These may actually be proved using DeMorgan's Theorem, as follows:

$$m_2 = \bar{A}B\bar{C}$$

$$m_2 = \overline{\overline{\bar{A}B\bar{C}}}$$

$$= \overline{\bar{\bar{A}} + \bar{B} + \bar{\bar{C}}}$$

$$= \overline{A + \bar{B} + C}$$

Therefore

$$m_2 = A + \bar{B} + C = M_5$$

In all previous functions each minterm has been specified as a 1 or a 0. However, many designs do not require all minterms to be defined, because the particular input conditions do not occur. For these situations the function may be a 0 *or* a 1. Thus it is unspecified and is called a "don't care." Later in this chapter it will be shown how the use of a "don't care" term can permit a function to be minimized further than if the term were not included.

Karnaugh Mapping

As has previously been demonstrated, one method of minimizing logic circuits is the use of Boolean algebra. But it is certainly not the only method, and in fact it does have a number of limits. In the circuits considered so far no mention of time or feedback paths has been made. Also, with other elements, such as counters and multivibrators, the output may depend either on the logic inputs at the time or on their state previously. Boolean algebra does not take this into account.

An alternative method of minimization that can be extended to include these items is that of Karnaugh mapping. This is an extensive method and therefore will only be dealt with in this chapter for circuits not possessing feedback. However, it should provide the reader with an alternative to Boolean algebra for the type of problem so far considered.

This method of minimization uses the basic truth-table sequence for functions as well as concepts from the Boolean algebra postulates and theorems. It is really a diagrammatic reduction of a statement rather than an algebraic one, and it is an extension of the Venn diagram principle.

The Karnaugh Map

The class K is defined as being the class of all points within a given rectangle or square; all variables are within a bounded area.

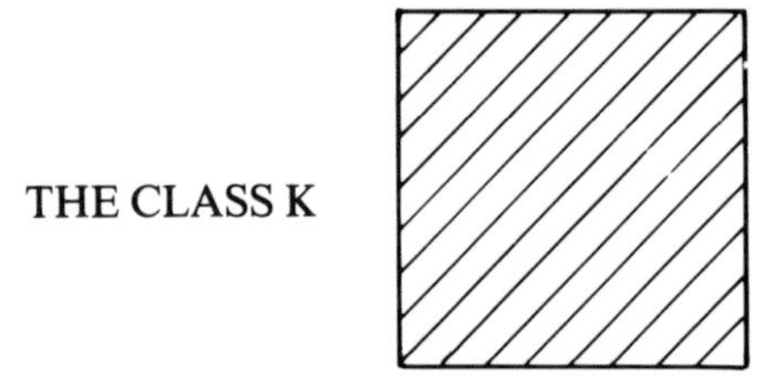

The Boolean variable A may then be represented by drawing a line across the square or rectangle so that the area below the line represents A and that above $\bar{A}$, as follows:

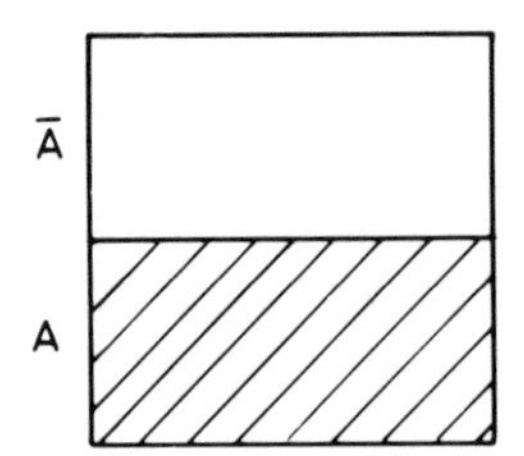

Thus the diagram is similar to that for Venn except that the areas are now blocks rather than circles.

This diagram or *map*, as it will hereafter be called, can then be extended to accommodate two variables, A and B, by drawing a vertical line down it. B is then defined to be the area to the right of the line and $\bar{B}$ the area to the left.

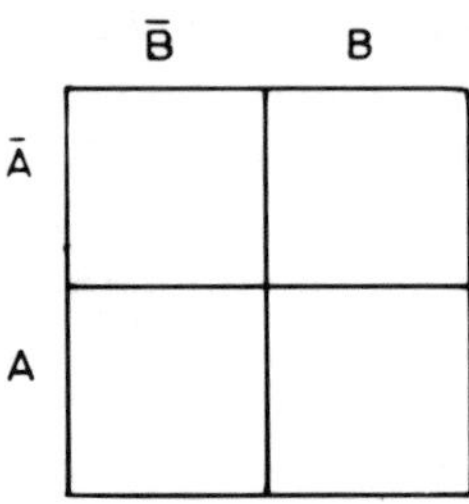

A two-variable Karnaugh map has now been generated. From this map the AND/OR functions can be interpreted.

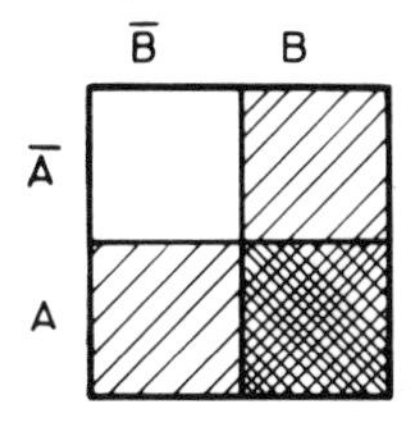

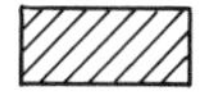

is the area covered by A OR B and therefore represents A + B.

is the area covered by both A AND B and therefore represents A · B.

Each square on the map is called a *cell,* and the two-variable Karnaugh map (K-map) has four cells.

An alternative way of drawing this map is to use the corresponding logic states in place of B and $\overline{B}$, as shown in figure 3–4, and inserting 0s and 1s in the corresponding cell position. The O/P is only a 1 when both A AND B are 1.

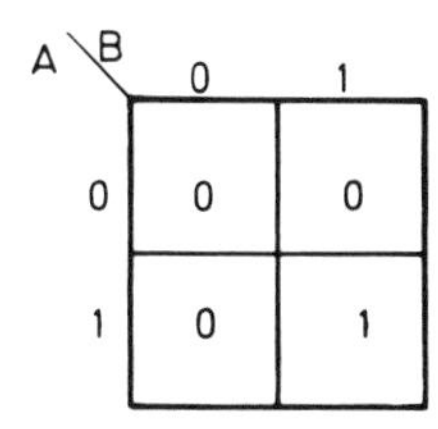

Figure 3–4. A Two-Variable Map for the AND Function

Three Variables

Figure 3–5 shows a three-variable map. The binary numbers along the top of the map indicate the condition of B and C for each column.

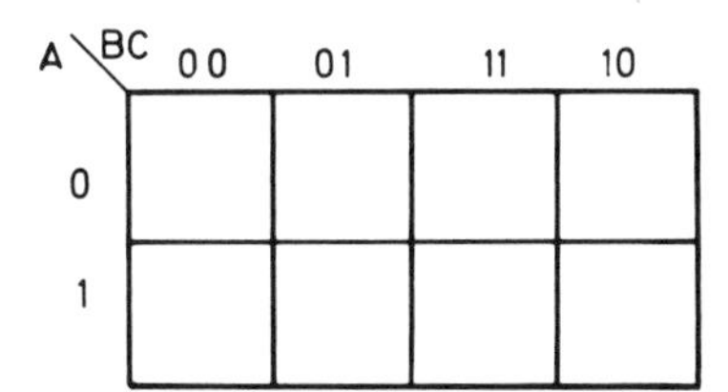

Figure 3–5. A Three-Variable Map

Once again the functions state for each input may be inserted into each cell position. For example, consider the following truth table which defines a particular function and the accompanying K-map which has been drawn for it.

A	B	C	*f*
0	0	0	0
0	0	1	0
0	1	0	0
0	1	1	1
1	0	0	0
1	0	1	1
1	1	0	0
1	1	1	1

The functions expression may be written as $f = \bar{A}BC + A\bar{B}C + ABC$, and therefore the K-map is actually a plot of the canonical sum-of-product form of the function. (See figure 3–6.)

Thus the function is a 1 when A = 0 AND B = 1 AND C = 1; *or* when A = 1 AND B = 0 AND C = 1; *or* when A = 1 AND B = 1 AND C = 1.

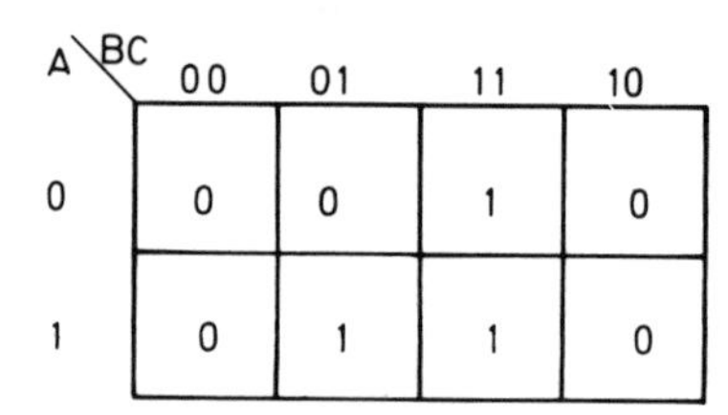

Figure 3–6. Karnaugh Map for the Function

Minimization Using K-Maps

For every K-map it is possible to label each cell by use of minterms (as defined previously in this chapter). A three-variable map is shown in figure 3–7, with the minterms labeled. (Note that the order of B and C along the

A \ BC	00	01	11	10
0	m_0	m_1	m_3	m_2
1	m_4	m_5	m_7	m_6

Figure 3–7. Minterm Positions on a K-Map

top of the map is not the regular binary sequence. This is because the whole theory of minimization using K-maps depends on the fact that only one variable is different between adjacent squares, as will be shown.)

It should be observed that the minterm number corresponds to the decimal equivalent of the binary number defining that cell. For example, 110_2 defines m_6 and 110 binary is equivalent to 6 decimal.

Consider the example given in the three-variable map section earlier in the chapter. The K-map for the function was:

A \ BC	00	01	11	10
0	0	0	1	0
1	0	1	1	0

The function is a logic 1 when m_3 OR m_5 OR m_7 is 1.
Therefore:

$$f = m_3 + m_5 + m_7$$

which may also be written as

$$f = \sum m(3, 5, 7)$$

or

$$f = \overline{A}BC + A\overline{B}C + ABC$$

Using the Boolean postulates and theorems, the function may be reduced as follows:

$$\begin{aligned} f &= \overline{A}BC + A\overline{B}C + ABC \\ &= \overline{A}BC + ABC + A\overline{B}C + ABC && \text{(by theorem 1)} \\ &= (\overline{A} + A) \cdot BC + AC \cdot (B + \overline{B}) && \text{(by postulate 7)} \\ &= BC + AC && \text{(by postulate 9)} \end{aligned}$$

The function has been reduced by combining m_5 ($A\overline{B}C$) and m_7 (ABC) to give AC and also by combining m_3 ($\overline{A}BC$) and m_7 (ABC) to give BC. This is the whole basis for minimization using K-maps: the fact that adjacent cells can be combined to reduce the expression and yet not change the function. It must be noted, however, that cells may only be combined in the horizontal or vertical directions because only then does the condition exist that one cell has one variable in it and the adjacent cell has the variable's complement (for example, m_5 and m_7, where B and $\overline{B}$ are the variable and its complement, respectively).

Additionally, it must be noted that cells occupying positions at opposite ends of the same row or column will also combine in this manner and that therefore the definition of adjacency must be extended to include these possibilities.

Examples of adjacency for a three-variable map are shown in figure 3-8. Cells that can be combined are "looped" on the K-map as demonstrated by 1, 2, 3, and 4.

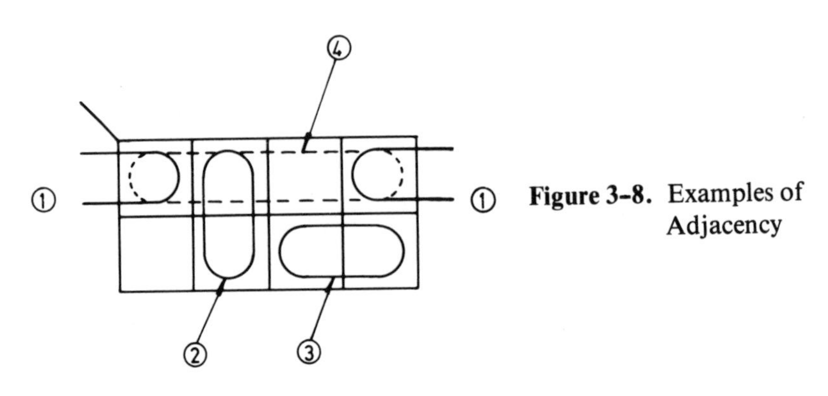

Figure 3-8. Examples of Adjacency

The overall rules for minimization using K-maps are, therefore:

1. Draw an appropriate map depending on the number of variables. When a function produces a logic 1, a 1 is placed in the corresponding cell of the K-map. Similarly, for a logic 0, a 0 is written.
2. Loop all those 1s that cannot combine with other 1s (that is, they are not adjacent to any other 1).
3. Loop all those 1s that will combine in a loop of 2 but will not combine in a loop of 4.
4. Loop all those 1s that will combine in 4s but not in 8s (in the case of maps with a larger number of variables than three, as will be shown) and so on, *such that each loop covers the maximum permissible area.*
5. Stop when all 1s have been covered.

Example 3.12

Minimize the function defined by the following truth table.

A	B	C	f
0	0	0	1
0	0	1	0
0	1	0	1
0	1	1	0
1	0	0	1
1	0	1	1
1	1	0	0
1	1	1	0

The function is

$$f = A\bar{B}C + \bar{A}B\bar{C} + A\bar{B}\bar{C} + \bar{A}\bar{B}\bar{C}$$

The K-map for it is shown below. All the 1s can combine with another 1 in a loop of 2, although none will combine in a loop greater than 2.

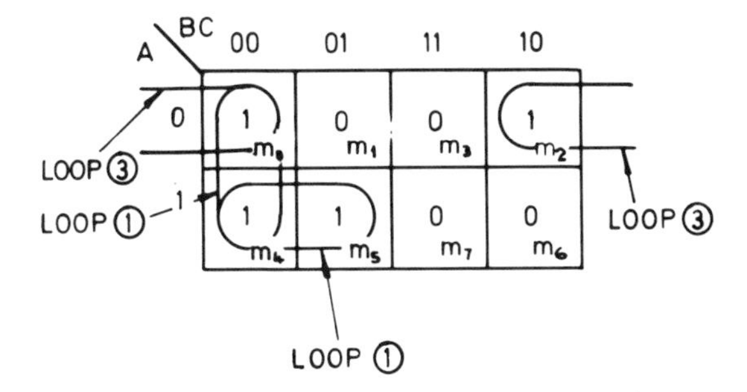

Note that both m_0 and m_4 have been covered twice. This is quite permissible if it is done to include an otherwise uncovered 1. To include it in other cases, however, will result in a redundant term in the answer.

Once the loops have been drawn the functions for them can be written down.

Loop 1 covers the area uniquely defined by B = 0 and C = 0. A can be either 0 or 1 and therefore does not define the loop. Hence loop 1 = $\bar{B}\bar{C}$.

Loop 2 covers the area uniquely defined by A being 1 and B being 0. In this case C can be either 0 or 1 and therefore does not define the loop. Hence loop 2 = $A\bar{B}$.

Loop 3 covers the area uniquely defined by A being 0 and C being 0. B is 0 or 1 for the loop and therefore does not define the loop. Hence loop 3 = $\overline{A}\overline{C}$.

The function is therefore a 1 when B = 0 AND C = 0; *or* when A = 1 AND B = 0; *or* when A = 0 AND C = 0. That is

$$f = \overline{B}\overline{C} + AB + \overline{A}\overline{C}$$

which is the minimized form for the function. The minimization has been achieved using Karnaugh mapping.

Example 3.13

Minimize the following function using K-maps.

$$f = \overline{A}\overline{B}C + \overline{A}B\overline{C} + A\overline{B}\overline{C} + A\overline{B}C$$

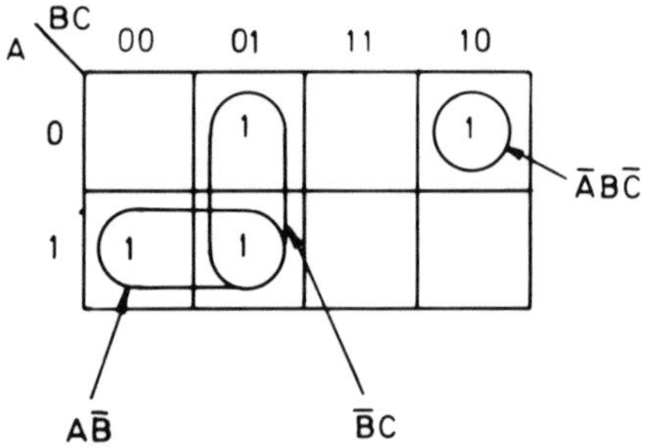

Thus the minimized function is

$$f = \overline{A}B\overline{C} + A\overline{B} + \overline{B}C$$

Four Variables

Figure 3-9 illustrates a four-variable map. Since there are four variables, there are sixteen possible states. These are labeled 0 through 15. Figures 3-10 and 3-11 show examples of cells that can be looped together if they contain a 1.

Five and Six Variables

For five-variable functions, two four-variable maps can be drawn. If the variables are ABCDE, then one map can be drawn for C and one for $\overline{C}$ as

Figure 3-9. A Four-Variable Map ▶

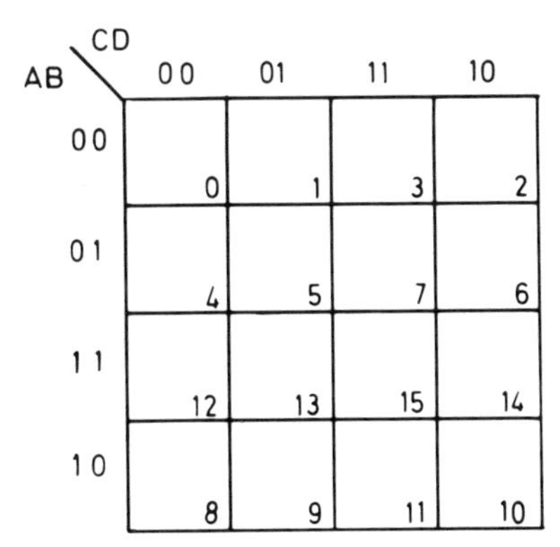

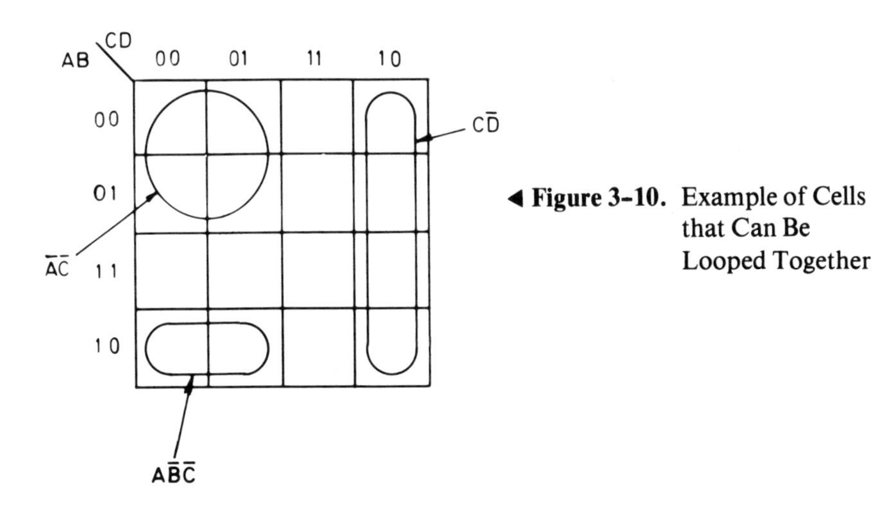

◀ **Figure 3-10.** Example of Cells that Can Be Looped Together

Figure 3-11. Example of Cells that Can Be Looped Together ▶

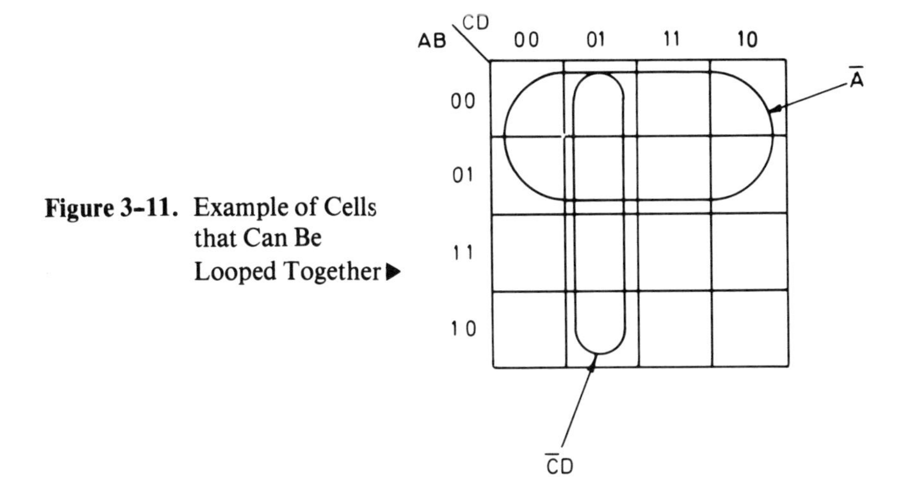

shown in figure 3–12. Similarly, a six-variable map can be drawn by drawing four four-variable maps, as shown in figure 3–13. If the variables in this case are ABCDEF, then the top two maps are for $\overline{A}$ and D or $\overline{D}$, whereas the bottom two are for A and D or $\overline{D}$.

On both figure 3–12 and figure 3–13 examples of adjacent cells have been given. If the five-variable map is labeled as shown, then adjacent cells occupy the same geographical positions. However if a five-variable map was constructed using a four-variable map for $\overline{E}$ instead of $\overline{C}$ and another

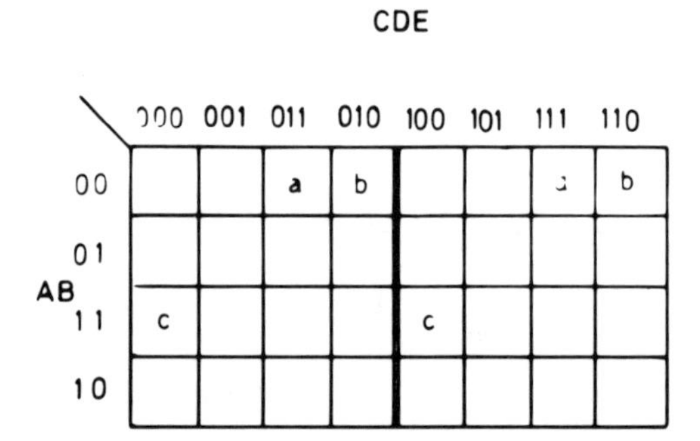

Figure 3–12. A Five-Variable Map

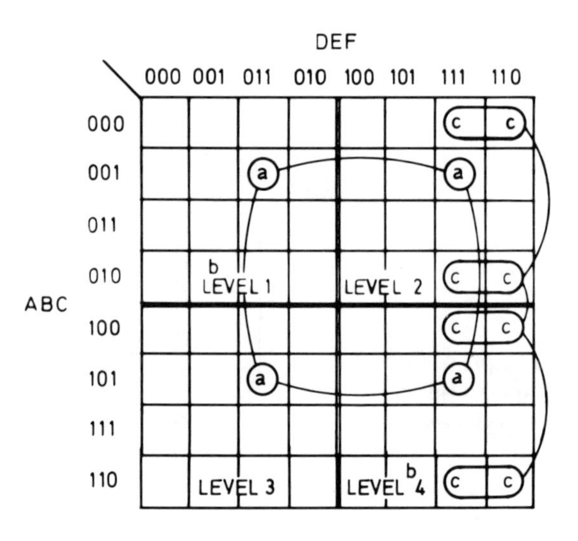

Figure 3–13. A Six-Variable Map

four-variable map for E instead of C, then adjacent cells would *not* occupy the same geographical position.

With six-variable maps adjacent cells in different "levels" (a "level" being one four-variable map) of the K-map might occupy the same geographical position in two levels, but not in all four. The levels in which adjacent cells exist must themselves still be adjacent according to the definition of adjacency.

This is exemplified in the six-variable map in that cells "a" are adjacent but not cells "b."

Cells "a" are all adjacent and so are cells "c," but cells "b" are *not* adjacent, that is, they have more than one variable different.

Example 3.14

Given the following truth table for a four-variable function, minimize the function using Karnaugh-mapping techniques.

A	B	C	D	*f*
0	0	0	0	1
0	0	0	1	1
0	0	1	0	0
0	0	1	1	1
0	1	0	0	0
0	1	0	1	0
0	1	1	0	0
0	1	1	1	1
1	0	0	0	0
1	0	0	1	0
1	0	1	0	0
1	0	1	1	0
1	1	0	0	0
1	1	0	1	1
1	1	1	0	0
1	1	1	1	1

The function *f* can then be written:

$$f = \bar{A}\bar{B}\bar{C}\bar{D} + \bar{A}\bar{B}\bar{C}\bar{D} + \bar{A}\bar{B}CD + \bar{A}BCD + ABCD + AB\bar{C}D$$

This function can now be plotted on a four-variable K-map and the function expression minimized using the adjacency theorem proved previously:

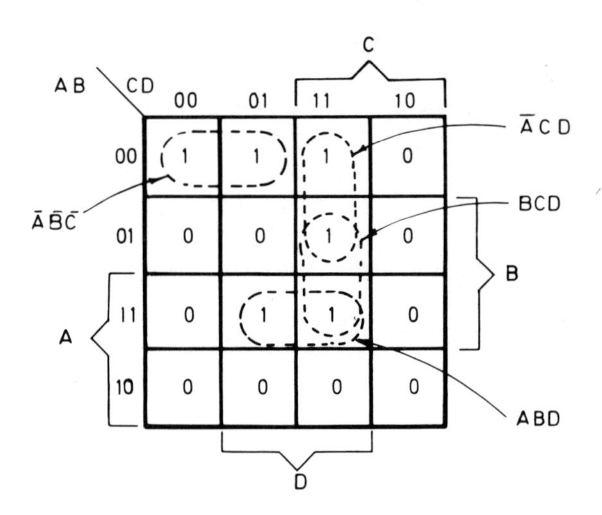

Therefore, $f = \overline{A}\overline{B}\overline{C} + BCD + ABD + \overline{A}CD$, which employs fewer terms and therefore fewer logic gates when a circuit is designed to produce the function.

An alternative solution that is equally correct would be $f = \overline{A}\overline{B}\overline{C} + BCD + ABD + \overline{A}\overline{B}D$; that is, ACD has been replaced by looping m_3 and m_1 instead of m_3 and m_7.

Exercise 3.4

Solve the following problems using K-maps

1. Simplify $f = A + \overline{A}B$
2. Simplify $f = \overline{A} \cdot \overline{B} + A \cdot \overline{B} \cdot \overline{C} + \overline{A} \cdot B \cdot \overline{C} + B \cdot C$
3. Simplify $f = A \cdot B \cdot \overline{C} \cdot \overline{D} + \overline{A} \cdot B \cdot \overline{C} + \overline{A} \cdot B \cdot \overline{D} + \overline{A} \cdot \overline{B} \cdot \overline{C}$
4. Simplify the function given by the following truth table:

A	B	C	D	*f*
0	0	0	0	1
0	0	0	1	1
0	0	1	0	1
0	0	1	1	0
0	1	0	0	0
0	1	0	1	X
0	1	1	0	X
0	1	1	1	X
1	0	0	0	X
1	0	0	1	X

Truth table continued.

A	B	C	D	f
1	0	1	0	X
1	0	1	1	0
1	1	0	0	1
1	1	0	1	1
1	1	1	0	1
1	1	1	1	0

X = Don't care.

5. Simplify $f = C(A \cdot \overline{B} + A \cdot B \cdot \overline{D} + A \cdot \overline{B} \cdot \overline{D}) + A \cdot \overline{C}$
6. The output of a digital unit is a parallel representation of a 4-bit natural binary code. Design a circuit whose output is high when the output of the digital unit is 5 or greater.
7. The terms $\overline{A}BCD\overline{E}$ and $AB\overline{C}DE$ are don't-care terms in the following Boolean expression:

$$f = \overline{A}\overline{B}D\overline{E} + \overline{A}C\overline{D}E + A\overline{C}\overline{D}E + \overline{A}\overline{C}\overline{D} + ABD\overline{E} + A\overline{C}D\overline{E} + ACDE + A\overline{B}\overline{C}E$$

Express the function as a minimal sum of products, using don't-care terms if required.

8. Each person on a panel of four may vote "hit" or "miss" on a number of topics. As each topic is considered, it is necessary to indicate whether the topic is approved or rejected by the panel or whether opinion is equally divided. Design a circuit, using AND, NAND, and OR gates, that will carry out this function. (Hint: Design a circuit for individual outputs and then combine the circuits.)

The Quine-McCluskey Minimization Procedure

The Quine-McCluskey procedure is basically a tabular method of minimization. In this method individual terms of an expression are described by their equivalent binary numbers and are then grouped according to the number of 1s occurring in the numbers. Numbers that differ by one 1 are paired, and a dash is written in place of the 1 that differs. Thus a second table is formed, and the process is repeated until a set of prime implicants are formed. (A prime implicant is a term that cannot be simplified any further, and these terms are ORed and ANDed together to form the final function expression.)

This method of minimization is probably best demonstrated by the following example.

Consider the function

$$f = \bar{A}\bar{B}\bar{C}\bar{D} + \bar{A}B\bar{C}D + \bar{A}BC\bar{D} + \bar{A}BCD + A\bar{B}\bar{C}D + A\bar{B}C\bar{D} + AB\bar{C}D + ABC\bar{D} + ABCD$$

This can be written as a sum of minterms, as follows:

$$f(A, B, C, D) = \sum m(0, 5, 6, 7, 9, 10, 13, 14, 15)$$

and these in turn can be written in tabular form as shown in table 3-1.

Minterms which have only one bit different are now paired to form table 3-2.

Numbers with two 1s in them and those with three are paired, where possible, once again. If not all of them can be paired, they are prime implicants and will appear in the final expression.

Table 3-1
List of Functions Minterms

Minterm	*A*	*B*	*C*	*D*		
0	0	0	0	0		
5	0	1	0	1	(a)	All numbers with two 1s in
6	0	1	1	0		
9	1	0	0	1		
10	1	0	1	0		
7	0	1	1	1		All numbers with three 1s in
13	1	1	0	1		
14	1	1	1	0		
15	1	1	1	1		

Table 3-2
Paired Minterms from Table 3-1

Minterms Paired	*ABCD*		
5, 7	01-1		All numbers with two 1s in
5, 13	-101		
6, 7	011-		
6, 14	-110		
9, 13	1-01	(b)	
10, 14	1-10	(c)	
7, 15	-111		All numbers with three 1s in
13, 15	11-1		
14, 15	111-		

Table 3-3
Paired Minterms from Table 3-2

Minterms Paired	*ABCD*	
(5, 7), (13, 15)	-1-1	(d)
(6, 7), (14, 15)	-11-	(e)

Note that minterms m_0, m_9, and m_{10} do not appear in table 3-3. Thus m_0 in table 3-1 and the pairing of m_9, m_{13} and m_{10}, m_{14} in table 3-2 must all be prime implicants. The overall minimized function will therefore consist of (a) from table 3-1; (b) and (c) from table 3-2 and (d) and (e) from table 3-3. That is:

$$f = (a) + (b) + (c) + (d) + (e)$$
$$= \bar{A}B\bar{C}D + A\bar{C}D + AC\bar{D} + BD + BC$$

4 Number Systems

The Decimal System

The decimal system of numbers is a system based on ten symbols: 0, 1, 2, 3, 4, 5, 6, 7, 8, and 9. In this system, the numbers ten, eleven, twelve, thirteen, and all higher numbers are made up of a combination of the basic symbols. Thus the number ten is made up of one and zero, namely 10.

The decimal number 4,675.235 signifies 4 thousands, 6 hundreds, 7 tens, 5 units, 2 tenths, 3 hundredths and 5 thousandths. It may be written more concisely as follows:

$$4 \times 10^3 + 6 \times 10^2 + 7 \times 10^1 + 5 \times 10^0 + 2 \times 10^{-1} + 3 \times 10^{-2} + 5 \times 10^{-3}$$

Presumably the decimal system was originally adopted because humans have ten fingers. It is no coincidence that a child in his early days at school uses his fingers when doing arithmetic. Had man evolved with only four fingers on each hand, the number systems used would very likely have been based on the number 8.

The Octal System

A system based on the number 8 uses the digits 0, 1, 2, 3, 4, 5, 6, and 7. Any numbers larger than seven are then made up of a combination of these basic symbols, just as with any number larger than nine in the decimal system. Such a number system is said to have a base or radix of 8 and is called the *octal* system. The following table shows the relationship between numbers in the decimal and octal systems.

Decimal	*Octal*	*Decimal*	*Octal*
0	0	4	4
1	1	5	5
2	2	6	6
3	3	7	7

Decimal	*Octal*	*Decimal*	*Octal*
8	10	15	17
9	11	16	20
10	12	17	21
11	13	18	22
12	14	19	23
13	15	20	24
14	16	21	25

Decimal-to-Octal Conversion

Decimal numbers may be split into two parts: (1) the whole number or integral part; and (2) the fractional part, that is, the part following the decimal point. When converting from decimal to octal, the integral part must be dealt with first. It is successively divided by 8 and the remainder noted at each step until the "answer" to the division by 8 is 0. The fact that the remainder at any stage is 0 is just as important as if it had been any other value and should be noted.

The fractional part is then multiplied successively by 8 and the overflow or whole number noted at each step. The operation is complete when the 0 state is reached or when a sufficient number of octal places has been achieved.

Example 4.1

To convert $(4{,}037.875)_{10}$ to an octal number.

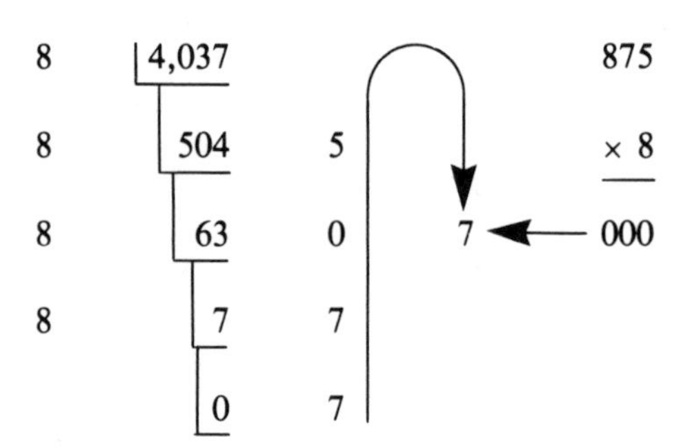

The result is read in the direction of the arrow.
Therefore, $(4{,}073.875)_{10} \equiv (7705.7)_8$.

Octal-to-Decimal Conversion

Octal, like decimal, is a positional number system; therefore, each digit has a particular weighting according to its position. Instead of weightings of 10^0, 10^1, 10^2, 10^3, and so on, for each column as for decimal, they are now 8^0, 8^1, 8^2, 8^3, and so on. Thus the octal number 4675.235 is actually:

$$(4 \times 8^3) + (6 \times 8^2) + (7 \times 8^1) + (5 \times 8^0) + (2 \times 8^{-1})$$
$$+ (3 \times 8^{-2}) + (5 \times 8^{-3})$$

By expanding this, the decimal equivalent of the octal number can be found:

$$(4 \times 512) + (6 \times 64) + (7 \times 8) + (5 \times 1) + \frac{2}{8} + \frac{3}{64} + \frac{5}{512}$$

$$= 2{,}048 + 384 + 56 + 5 + \frac{157}{512}$$

$$= \left(2{,}493\,\frac{157}{512}\right)_{10}$$

Therefore, $(4675.235)_8 \equiv \left(2{,}493\,\frac{157}{512}\right)_{10}$

An alternative method of converting the integer part of an octal number to decimal is to multiply the most significant digit of the octal number by 8 and add the next digit to the right to the result. Multiply the result by 8 and add the next digit to the result, and so on until the end of the number. This is demonstrated by converting 4675_8 to decimal.

Copy the left digit: 4

Multiply by 8 and add the next digit: $(8 \times 4) + 6 =$ 38

Multiply by 8 and add the next digit: $(8 \times 38) + 7 =$ 311

Multiply by 8 and add the next digit: $(8 \times 311) + 5 =$ 2,493

Therefore $4675_8 \equiv 2{,}493_{10}$.

To convert the fractional part of an octal number into decimal, the last bit must be divided by 8 and the next digit to the left added. The result is divided by 8, the next digit added and the result divided by 8, and so on until the last digit of the fractional part of the number has been considered. This is demonstrated by converting 0.235_8 to decimal, as follows:

Copy the last digit: 5

Divide by 8 and add next digit to the left: $5/8 + 3 = 29/8$

Divide by 8 and add next digit to the left: $29/64 + 2 = 157/64$

Divide by 8: $= 157/512$

This alternative method can in fact be used to convert any number system of a base other than 10 to decimal. Therefore, to convert a number of base *n*, the rule is to multiply by *n* and add the next digit, for the whole number portion, and to divide by *n* and add for the fractional part.

Exercise 4.1

1. Write down the decimal equivalent of the following:
   a. $(2367.56)_8$
   b. $(12031.102)_8$
   c. $(6421.28)_8$
2. a. Convert $(1{,}359.275)_{10}$ to octal (to 3 octal places).
   b. Convert $(2{,}382.125)_{10}$ to octal.

The Binary System

Decimal and octal systems are not the only possible number systems. Another even simpler system in use is the *binary* system of numbers, which has only two symbols: 0 and 1. The number 2 (decimal) does not have a symbol of its own but is written 10 (1 and 0).

Therefore, instead of column weighting of 10^0, 10^1, 10^2, 10^3, and so on as for decimal, they are now 2^0, 2^1, 2^2, 2^3, and so on.

Decimal-to-Binary Conversion

This is carried out in a manner similar to that for decimal to octal conversion, except that the integral part of the decimal number is now divided successively by 2 rather than by 8. Similarly, the fractional part is successively multiplied by 2 until the 0 state is reached.

Example 4.2

Convert $(248.625)_{10}$ to binary.

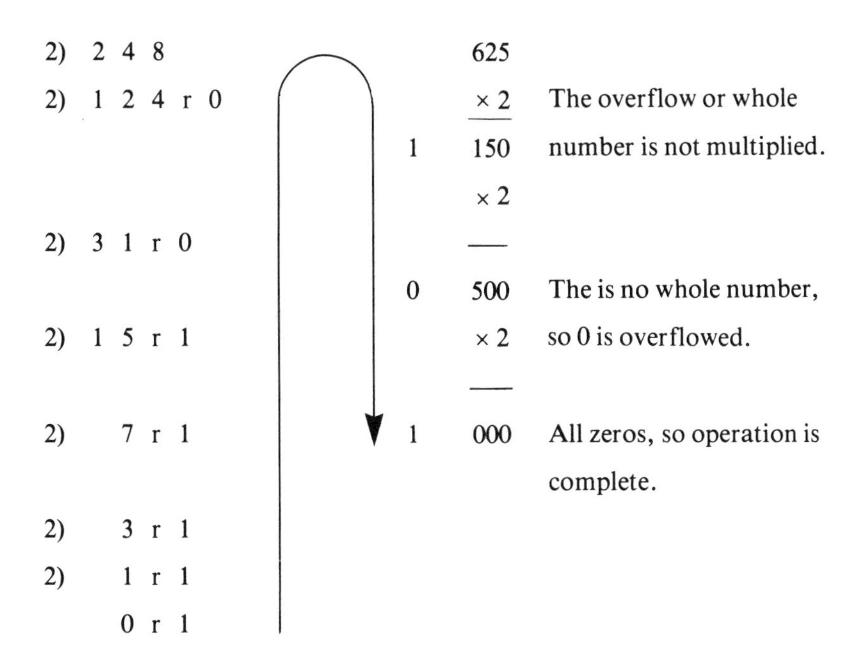

The binary number is then read off in the direction of the arrow. Thus:

$$(248.625)_{10} \equiv (11111000.101)_2$$

Note that there are some fractions that cannot be exactly specified in binary, in the same way that 1/3 cannot be exactly specified in decimal.

Example 4.3

Convert $(4{,}037.875)_{10}$ to binary and check the answer.

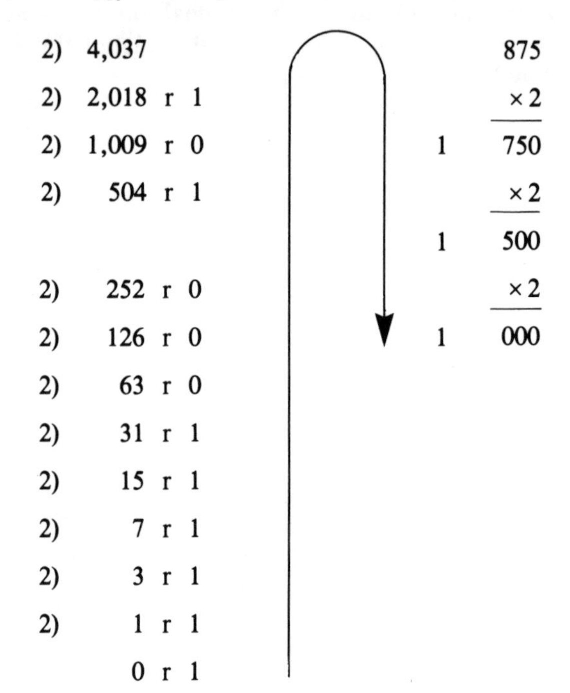

Therefore $(4{,}037.875)_{10} = (111111000101.111)_2$

Check:

$$111111000101.111 =$$

2048	1024	512	256	128	64	32	16	8	4	2	1 ·	$\frac{1}{2}$	$\frac{1}{4}$	$\frac{1}{8}$
1	1	1	1	1	1	0	0	0	1	0	1 ·	1	1	1

$$\begin{aligned} &= 2048 \\ &+ 1024 \\ &+ 512 \\ &+ 256 \\ &+ 128 \\ &+ 64 \\ &+ 4 \\ &+ 1 \\ &\overline{\quad 4037} \end{aligned}$$

$$4037\frac{7}{8} = (4037.875)_{10}$$

Therefore, the conversion from decimal to binary was correct.

An alternative method of converting the integer part of a binary number to decimal is to multiply the most significant bit of the binary by 2 and add the next digit to the right to the result. Multiply the result by 2 and add the next digit to the result and so on until the end of the number. This is demonstrated by converting the binary number (101101.101) of example 4.3 using this method.

Copy the left bit: 1

Multiply by 2 and add the next bit: $(2 \times 1) + 0 = 2$

Multiply by 2 and add the next bit: $(2 \times 2) + 1 = 5$

Multiply by 2 and add the next bit: $(2 \times 5) + 1 = 11$

Multiply by 2 and add the next bit: $(2 \times 11) + 0 = 22$

Multiply by 2 and add the next bit: $(2 \times 22) + 1 = 45$

Therefore $101101_2 \equiv 45_{10}$.

To convert the fractional part of the number, the last digit must be divided by 2 and the next digit to the left added. The result is divided by 2 and the next digit added and so on up to the last digit of the fractional part of the number. The solution for 0.101_2 is:

Copy the first number: 1

Divide by 2 and add the next digit: $1/2 + 0 = 1/2$

Divide by 2 and add the next digit: $1/4 + 1 = 5/4$

Divide result by two: $= 5/8$

Exercise 4.2

Convert the following numbers to binary:

1. $(23)_{10}$
2. $(100)_{10}$
3. $(145)_{10}$
4. $(183.65)_{10}$
5. $(1{,}083.6875)_{10}$
6. $(3{,}047.827)_{10}$

Binary-to-Decimal Conversion

As mentioned previously, the columns in the binary system represent power of 2. Therefore, a binary number may be converted to decimal by multiplying each binary digit by its respective column power and adding them all together, as shown in the next example.

Example 4.4

Convert the binary number $(101101.101)_2$ to decimal.

To make conversions from binary to decimal more easily seen, the column headings may be written in their decimal equivalents, as follows:

$$\begin{array}{cccccccccc} 32 & 16 & 8 & 4 & 2 & 1 & \cdot & \frac{1}{2} & \frac{1}{4} & \frac{1}{8} \\ 1 & 0 & 1 & 1 & 0 & 1 & \cdot & 1 & 0 & 1 \end{array}$$

Therefore the decimal equivalent is:

$$32 + 8 + 4 + 1 + \frac{1}{2} + \frac{1}{8} = (45\tfrac{5}{8})_{10}$$

Exercise 4.3

Convert the following numbers to decimal:

1. $(1111.111)_2$
2. $(10000.001)_2$
3. $(10101101.101)_2$

Octal-to-Binary Conversion

The following table shows how the eight symbols used in the octal system may be represented by a 3-bit binary number.

Octal	*Binary*	*Octal*	*Binary*
0	000	4	100
1	001	5	101
2	010	6	110
3	011	7	111

To convert an octal number into binary, each octal digit is simply replaced by its three-digit binary equivalent. Thus:

$$(7705.7)_8 = \begin{matrix} 7 & 7 & 0 & 5 & 7 \\ 111 & 111 & 000 & 101 & 111 \end{matrix}$$

$$(7705.7)_8 \equiv (111111000101.111)_2$$

Decimal-to-octal conversion is far less tedious than decimal-to-binary. Since octal-to-binary conversion is such a simple matter, a quicker way of carrying out decimal-to-binary conversion is to convert to octal and then to binary. This is demonstrated by the following example.

Example 4.5

Convert $(3{,}935.65625)_{10}$ to octal and then to binary.

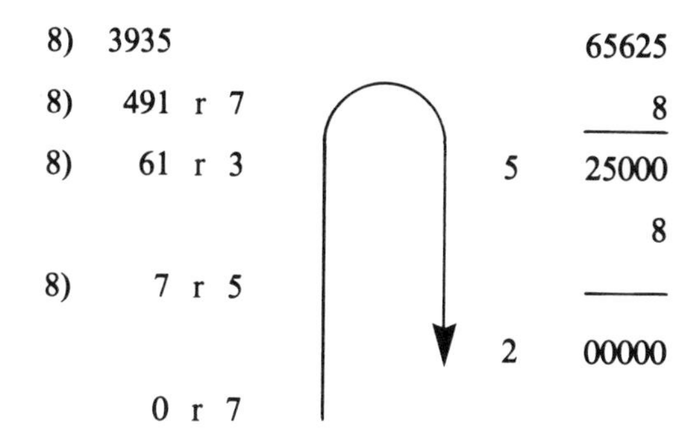

Therefore:

$$(3{,}935.65625)_{10} = (\,7 \quad 5 \quad 3 \quad 7 \quad 5 \quad 2\,)_8$$
$$= (111 \quad 101 \quad 011 \quad 111\,.\,101 \quad 010)_2$$

The answer may be checked by converting the binary number back to decimal. In general, when converting from decimal to binary it is best to put in an intermediate step of conversion to octal, since this not only saves time but also reduces the chance of arithmetic error by requiring fewer arithmetical steps.

Binary-to-Octal Conversion

This is simply the reverse process of octal-to-binary. The binary number is split into groups of three digits, starting at the least-significant end (right-hand side); then the corresponding octal number for each group is written down.

Example 4.6

Convert $(1101001.111)_2$ to octal.
Therefore:

$$(1101001111)_2 = (1517)_8$$

Exercise 4.4

1. Convert the following octal numbers to binary:
   a. $(2610)_8$
   b. $(742)_8$
   c. $(6003)_8$
2. Convert the following binary numbers to octal:
   a. 1101110_2
   b. 11110011_2
   c. 010111001_2

Binary Arithmetic

Arithmetic functions can be carried out in a similar way for binary and decimal, with the notation for the various quantities remaining the same.

Augend + addend	= Sum
Minuend − subtrahend	= Difference
Multiplicand (Md) × multiplier (Mr)	= Product
Dividend (Dd) ÷ divisor (Dr)	= Quotient

Binary Addition

When adding two decimal numbers together, each column is dealt with separately, starting from the least-significant end. This is because each column is being examined to see whether its total generates a carry to be added to the next column of greater significance.

Example 4.7

Add 42 to 64 in decimal:

Augend	64
Addend	42
Carry	100
Sum	106

The addition of 4 and 2 in the 10^0 column produces 6×10^0 with no carry. However, the sum of 6×10^1 and 4×10^1 produces 10×10^1, or 10^2. 10^2 is equal to $(1 \times 10^2) + (0 \times 10^1)$, and therefore a 0 was written in the 10^1 column and a 1 was carried into the 10^2. A similar procedure is used in binary addition, except that the only symbols are 0 and 1. The rules for binary addition are as follows:

$$0 + 0 = 0$$

$$1 + 1 = 1$$

$$0 + 1 = 1$$

$$1 + 1 = 0 \text{ with 1 to carry}$$

Example 4.8

$$0101101 \equiv (1 \times 2^5) + (1 \times 2^3) + (1 \times 2^2) + (1 \times 2^0) \equiv 45$$

$$\underline{0011100} \equiv \qquad (1 \times 2^4) + (1 \times 2^3) + (1 \times 2^2) \equiv 28$$

$$1001001 \equiv (1 \times 2^6) + (1 \times 2^3) + (1 \times 2^0) \equiv 73$$

The same method is used for any addition problem, whether the numbers used are mixed or otherwise.

$$010111.11 \equiv 23.75$$

$$\underline{001011.01} \equiv 11.25$$

$$100011.00 \equiv 35.00$$

Binary Subtraction

In decimal subtraction the method is similar.

Example 4.9

Subtract 84 from 106.

Minuend	106
Subtrahend	84
Borrow	100
Difference	22

The subtraction of 4 from 6 is not difficult, with the result being placed in the 10^0 column. However, 8 cannot be subtracted from 0, and therefore a borrow from the 10^2 column must be made. The subtraction now becomes $10 - 8 = 2$. The borrowed 1 from the 10^2 column must now be paid back. Then, considering the 10^2 column, we have $1 - 1 = 0$.

The rules of binary subtraction are as follows:

$$0 - 0 = 0$$

$$0 - 1 = \text{ and borrow } 1$$

$$1 - 1 = 0$$

$$1 - 0 = 1$$

Example 4.10

Subtract 011100 from 101101.

$$101101 = 45$$

$$\underline{011100} = 28$$

$$010001 = 17$$

Binary Multiplication

The four rules for binary multiplication are as follows:

$$0 \times 0 = 0$$

$$0 \times 1 = 0$$

$$1 \times 0 = 0$$

$$1 \times 1 = 1$$

The principles of binary multiplication are similar to those for decimal numbers, although the simplicity of the binary rules means that the formation of each partial product is achieved by multiplying the multiplicand by either 0 or 1.

Example 4.11

Multiply 1101 by 1011, that is, 13_{10} by 11_{10}.

```
       1101     Md
       1011     Mr
       ----
       1101     Md × (1 × 2^0)
      1101      Md × (1 × 2^1)
     0000       Md × (0 × 2^2)
    1101        Md × (1 × 2^3)
    --------
    10001111    Product = 143
```

Note that as in decimal long multiplication there are as many partial products as there are digits in the multiplier, and the addition of these partial products produces the final product.

Binary Division

In its simplest form, division can be achieved by a series of subtractions in which the divisor is repeatedly subtracted from the dividend until the remainder is smaller than the divisor. The quotient is equal to the number of subtractions.

Example 4.12

Divide 12 by 3.

$12 - 3 = 9$	Partial quotient $= 1$
$9 - 3 = 6$	$= 2$
$6 - 3 = 3$	$= 3$
$3 - 3 = 0$	Quotient $= 4$

This method is obviously slow when the quotient is large.

Binary division can also be carried using a method similar to that for long division with decimal numbers. The process is basically a series of trial-and-error movements, but, as with multiplication, the binary method is simpler than the decimal one.

Example 4.13

Divide 10010000 by 1100, that is, 144_{10} by 12_{10}.

```
           1100
     ----------
 1100)10010000
       1100
       ----
      001100
        1100
        ----
      000000
```

The divisor will divide into the most significant five dividend digits once, and a 1 is placed above the fifth dividend digit. The divisor is subtracted from these dividend digits, and the sixth dividend digit is added to the partial remainder. The process continues until the quotient is formed.

Exercise 4.5

1. Find the sum of 011011_2, 001011_2, and 010111_2. Check the result by conversion to decimal.
2. Find the sum of 100111.011_2 and 011010.101_2. Check the result by conversion to decimal.
3. Subtract 10101_2 from 110010_2. Check the result by conversion to decimal.
4. Subtract 001011.01_2 from 10101.11_2. Check your result.
5. Multiply:
   a. 11001_2 by 1101_2.
   b. 101110_2 by 110_2.
6. Divide:
   a. 110111000_2 by 1011_2.
   b. 101010010_2 by 1101_2.

Binary-Coded-Decimal Number System

An outstanding disadvantage of binary and octal numbers is that their conversion to decimal becomes more cumbersome as the length of the number increases. There is no repeating conversion pattern like the one that exists between octal and binary numbers. This limitation is most inconvenient at the input from binary equipment, since the normal source and result are numbers expressed in decimal form. In some cases it is convenient to use a form of coding known as binary-coded decimal (bcd), in which each decimal digit is converted into a form of binary coding independent of the remaining decimal digits.

8, 4, 2, 1 Binary-Coded-Decimal

This is one form of binary-coded-decimal code in which the digits 0–9 are allocated the following binary codings.

0	0000
1	0001
2	0010
3	0011
4	0100
5	0101
6	0110
7	0111
8	1000
9	1001

This is of course the normal pure binary coding, but the remaining six codes are not used. In this sense the binary-coded-decimal system is wasteful, since it uses four binary digits to account for *each* decimal digit.

Example 4.14

Express 654 in 8, 4, 2, 1 bcd.

$$\overbrace{0110}^{6} \quad \overbrace{0101}^{5} \quad \overbrace{0100}^{4} \text{ bcd}$$

Two other forms of bcd coding exist, but in both of these there is a possible ambiguity in coding. One of these is 4, 2, 2, 1 bcd. In this system decimal 9 is coded as 1111, but decimal 5 may be 1001 or 0111. Similarly, decimal 3 may be 0101 or 0011. In practice, coding from the least-significant *end* to the most-significant *end* resolves the ambiguity.

The second alternative form of bcd coding is 4, 4, 2, 1 bcd, which is similar to 4, 2, 2, 1 bcd.

Although the various bcd codes are convenient with respect to conversion to and from decimal, bcd numbers cannot be used in normal arithmetic operations and would require conversion to pure binary if such operations were required with data supplied as bcd. In this case it may be better to use pure binary throughout. In fact it is possible to design arithmetic units for bcd numbers instead of pure binary numbers, but this is rarely done.

Hexadecimal Number System

Hexadecimal was designed, like the octal number system, to meet the need of expressing binary numbers more concisely. Many computer word lengths are 8 bits, 16 bits, or 32 bits; therefore, grouping in 4-bit groups instead of 3-bit groups is an improvement. This means that each group of 4 bits can define 16 unique numbers, 0 through 15_{10}. However, there are no symbols within our number system that represent the numbers 10_{10} to 15_{10} using one symbol only. Therefore, it was decided to use the letters A, B, C, D, E, and F to represent these numbers, as shown in the following table.

Hexadecimal is a number system of base 16, and therefore the number 16 decimal is equivalent to 10 (one-zero) hexademical.

Decimal	*Binary*	*Hexadecimal*
0	0000	0
1	0001	1
2	0010	2
3	0011	3
4	0100	4
5	0101	5
6	0110	6
7	0111	7
8	1000	8
9	1001	9
10	1010	A
11	1011	B
12	1100	C
13	1101	D
14	1110	E
15	1111	F
16	10000	10
17	10001	11

Conversion of Binary to Hexadecimal

Binary numbers are converted to hexadecimal by dividing up the binary number into groups of 4 bits, beginning with the least significant bit, and writing down the corresponding hexadecimal number for each group. This is demonstrated by the following example.

Example 4.15

Convert 1011001011000110_2 to hexadecimal.

1011	0010	1100	0110_2
B	2	C	6

Therefore $1011001011000110_2 = B2C6_{16}$

Note that hexadecimal-to-binary conversion is simply the reverse process.

Conversion of Decimal to Hexadecimal

To convert a decimal number to hexadecimal the number must be successively divided by 16 and the remainder noted. It must be remembered that any remainder greater than nine must be written in the corresponding hexadecimal format A to F.

Example 4.16

Convert 302_{10} to hexadecimal.

```
16 ) 302
16 )  18   r E
16 )   1   r 2
       0   r 1
```

Therefore $302_{10} \equiv 12E_{16}$

Conversion of Hexadecimal to Decimal

As for other number systems, conversion from hexadecimal to decimal may be achieved by multiplying each digit by its respective column weighting. In this case the column weightings will be 16^0, 16^1, 16^2, 16^3, and so on for the whole-number part, and 16^{-1}, 16^{-2}, 16^{-3}, and so on for any fractional part.

Example 4.17

Convert $12E.3D_{16}$ to decimal.

$$
\begin{aligned}
E \times 16^0 &= 14 \times 1 = 14 \\
2 \times 16^1 &= 2 \times 16 = 32 \\
1 \times 16^2 &= 1 \times 256 = \underline{256} \\
& \qquad\qquad 302
\end{aligned}
$$

$$
\begin{aligned}
3 \times 16^{-1} &= 3 \times 1/16 = 3/16 \\
D \times 16^{-2} &= 13 \times 1/256 = \underline{13/256} \\
& \qquad\qquad 61/256
\end{aligned}
$$

Therefore $12E.3D_{16} \equiv (302\frac{61}{256})_{10}$

As shown previously for binary and octal, an alternative method of conversion is to multiply each digit by the base and add for the whole-number part and to divide each digit by the base and add (for the fractional part). Therefore, for conversion from hexadecimal to decimal, the factor used for the multiplication and division is 16. This is demonstrated by converting the previous example using this method.

Solution

Whole-Number Part

Copy down the first digit: 1

Multiply by 16 and add the next digit: $(16 \times 1) + 2 = 18$

Multiply by 16 and add the next digit: $(16 \times 18) + 14 = 302$

Fractional Part

Copy down first number: 13

Divide by 16 and add next digit: $13/16 + 3 = 61/16$

Divide by 16: $= 16/256$

Therefore $12E.3D_{16} \equiv (302\frac{61}{256})_{10}$

Table 4-1
Hexadecimal Addition Table

	0	*1*	*2*	*3*	*4*	*5*	*6*	*7*	*8*	*9*	*A*	*B*	*C*	*D*	*E*	*F*
0	0	1	2	3	4	5	6	7	8	9	A	B	C	D	E	F
1	1	2	3	4	5	6	7	8	9	A	B	C	D	E	F	10
2	2	3	4	5	6	7	8	9	A	B	C	D	E	F	10	11
3	3	4	5	6	7	8	9	A	B	C	D	E	F	10	11	12
4	4	5	6	7	8	9	A	B	C	D	E	F	10	11	12	13
5	5	6	7	8	9	A	B	C	D	E	F	10	11	12	13	14
6	6	7	8	9	A	B	C	D	E	F	10	11	12	13	14	15
7	7	8	9	A	B	C	D	E	F	10	11	12	13	14	15	16
8	8	9	A	B	C	D	E	F	10	11	12	13	14	15	16	17
9	9	A	B	C	D	E	F	10	11	12	13	14	15	16	17	18
A	A	B	C	D	E	F	10	11	12	13	14	15	16	17	18	19
B	B	C	D	E	F	10	11	12	13	14	15	16	17	18	19	1A
C	C	D	E	F	10	11	12	13	14	15	16	17	18	19	1A	1B
D	D	E	F	10	11	12	13	14	15	16	17	18	19	1A	1B	1C
E	E	F	10	11	12	13	14	15	16	17	18	19	1A	1B	1C	1D
F	F	10	11	12	13	14	15	16	17	18	19	1A	1B	1C	1D	1E

Hexadecimal Addition and Subtraction

Hexadecimal addition and subtraction is performed in a way similar to that for decimal operations. A hexadecimal addition table such as the one shown in table 4-1 is quite useful until experience in addition has been obtained.

Example 4.18

Add $1B6_{16}$ and $48E_{16}$.

Columns

X	Y	Z
1	B	6
+4	8	E

Column X:

$$6 + E = 6 + 14_{10}$$
$$= 20_{10}$$

$$= 16_{10} + 4$$

$$= 14_{16}$$

Sum of 4, carry of 1.

Column Y:

$$1 + B + 8 = 1 + 11_{10} + 8$$

$$= 20_{10}$$

$$= 16_{10} + 4$$

$$= 14_{16}$$

Sum of 4, carry of 1.

Column Z:

$$1 + 1 + 4 = 6$$

Sum of 6, no carry

Thus $1B6_{16} + 48E_{16} = 644_{16}$.

Subtraction can also be carried out using the addition table. To use the table, find the smaller of the two numbers along the left edge. Then move horizontally along the row until the higher of the two numbers is found. The result may then be read from the top of this row.

For example, find C − 8. First find 8 along the left edge, and then move across the row until C is found. The answer—4—then appears at the top of the column.

Example 4.19

Subtract 3B2 from 591.

	Column	
X	Y	Z
5	9	1
−3	B	2

Column X: Borrow 1 from column Y.

$$11 - 2 = F$$

Column Y: B now becomes C as borrow for column Z.

Borrow 1 from column C.

19 − C = D

Column Z: 3 becomes 4 as borrow for column Y.

5 − 4 = 1

Thus $591_{16} - 3B2_{16} = 1DF_{16}$

Exercise 4.6

1. Convert the following to 8, 4, 2, 1 bcd:
   a. 385_{10}
   b. 926_{10}
2. Convert the following to hexadecimal:
   a. 1101010110100011_2
   b. 0011100111110110_2
   c. 11010111010111_2
3. Add the following hexadecimal numbers:
   a. 38 + AC
   b. 7DE1 + A786
4. Subtract the following hexadecimal numbers:
   a. 1A from 39
   b. 1A9 from 7683

Excess-3 Code

The excess-3 code (XS3) derives its name from the fact that each code is the 8, 4, 2, 1 binary code plus 0011 (3). Thus:

0_{10} is 0000 + 0011 = 0011,

1 is 0001 + 0011 = 0100,

2 is 0010 + 0011 = 0101, and so on.

When adding XS3, the binary numbers are added normally; then, if there is no carry out from the 4-bit group, 0011 must be subtracted. If a carry out occurs, 0011 must be added.

Example 4.20

Add 3 and 2 in XS3.

3 =	0110
2 =	0101
Sum	1011
No carry ∴ subtract	0011
	1000 = 5

Example 4.21

Add 7 and 8 in XS3.

7 =	1010
8 =	1011
Sum	10101
Carry out ∴ Add	0011
15 =	11000

Exercise 4.7

1. Add 2 and 5 in XS3.
2. Add 4 and 6 in XS3.

Computer-Arithmetic—One's and Two's Complement

The truth table for binary subtraction may be used to design a *binary subtractor,* but in general this is not done. In a computer where it is desired to subtract two numbers, the internal representation of the numbers is modified to take account of negative numbers, and subtraction is then the addition of a negative number. Thus no further hardware other than an adder is required.

The format of binary numbers is changed, and a sign bit is introduced in the most-significant bit position. The usual convention is:

If MSB = 0, then the number is positive.

If MSB = 1, then the number is negative.

The two modified binary formats are called one's-complement and two's-complement representation.

Signed One's-Complement Representation

Numerically the one's complement of an n-bit number X is $2^n - X - 1$. It is obtained by complementing the whole number.

$+5_{10} = 0\,101_2$

sign bit (arrows pointing to the leading 0 above and the leading 1 below)

$-5 = 1\,010\ (2^3 - 5 - 1 = 2)$ 2 in binary, ignoring leading 1

Subtraction Using One's Complement to Perform $(Y - X)$

1. One's complement X: $2^n - X - 1$
2. Add Y.
3. Subtract using $Y +$ one's complement of X.
4. $\therefore$ Condition for carry out is:

$$Y + 2^n - X - 1 > 2^n$$

That is

$$2^n + Y - X - 1 > 2^n$$

$$Y - X > 1$$

5. $\therefore X < Y$.

For $X < Y$ there is a carry from the 2^{n-1} to the 2^n stage; and since the 2^n stage always has a 1, there is a carry out from this stage. The carry-out 1 is therefore added to the least-significant state (EAC = end-around carry)

If $X \geq Y$, one's complement result:

$$2^n - \underbrace{(2^n - (X - Y) - 1) - 1)}_{\text{result} - 1} = X - Y$$

$$2^n - \quad \text{result} - 1$$

Example 4.22

Find 7 − 5.

7_{10}	0111
5_{10}	0101
7	0111
− 5	1010
	0001
EAC	+ 1
	0010 = +2

Example 4.23

Find 2 − 4.

2	0010	0010	
4	0100	1011	
		1101	No carry ∴ one's complement
gives		0010 = 2	
∴ Answer		= −2	

The disadvantage of one's complement is that there are two forms for 0:

$$1111 = -0$$

$$0000 = +0$$

Signed Two's-Complement System

The two's complement of an n-bit number is $2^n - X$. Therefore, to convert to two's complement representation, one's complement the number and then add 1:

$$X \longrightarrow 2^n - X - 1 \longrightarrow 2^n - X$$

Subtraction: $Y - X$

1. Two's complement X
2. Add Y
   a. If carry out, then read result directly.
   b. If no carry out, then two's complement result, that is, answer is negative.

Example 4.24

Find 7 − 5.

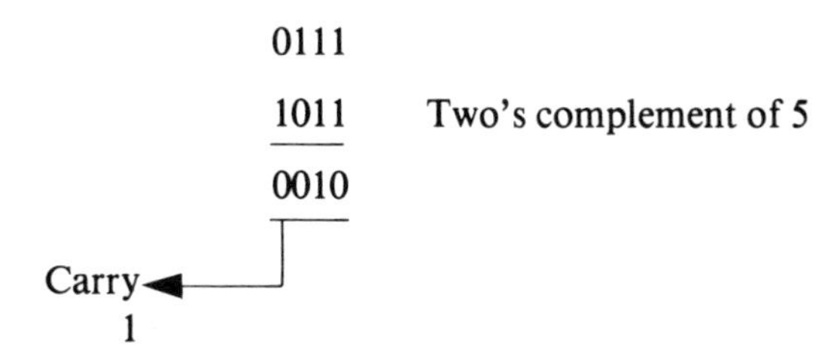

The MSB is 0 and hence the sign is positive.

Answer = +2

Example 4.25

Find 2 − 4.

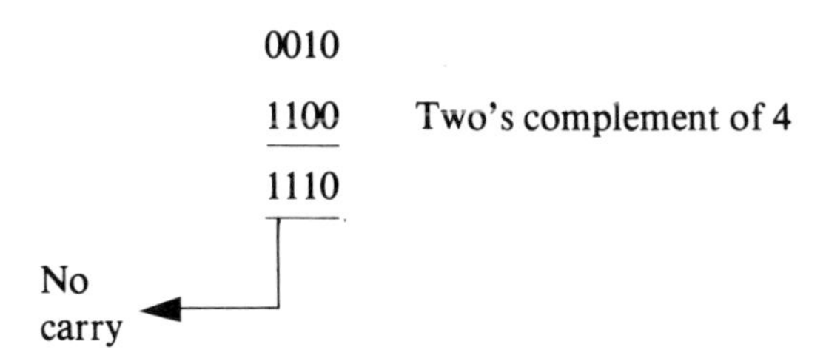

The MSB is 1 and hence the sign is negative.

Answer is two's complement of 1100

= −2

Note that with larger numbers a simpler way of converting to two's complement is to invert the number and add 1.

Example 4.26

Find the two's complement of 345_{16}

$$345_{16} = 0011\ 0100\ 0101_2$$

Inverting this gives 110010111010
+ 1

110010111011 = CBB_{16}

Therefore, the overall process is:

251_{16}	=	001001010001_2	
Inverted	=	110110101110	
+ 345_{16}	=	001101000101	
1		000011110011	
		+ 1	ADDIN CARRY
1		000011110100 = 0F4	

With this result a carry is generated; therefore, following the rule for two's complement arithmetic, the result is read directly, that is, + 0F4.

If no carry had been generated, then the output would have had to be two's complemented to obtain the result. This could be achieved by returning the output to the B-gate input, inverting it, applying 0 to the A-gate input, and generating ADDIN CARRY. This is demonstrated by the following diagram.

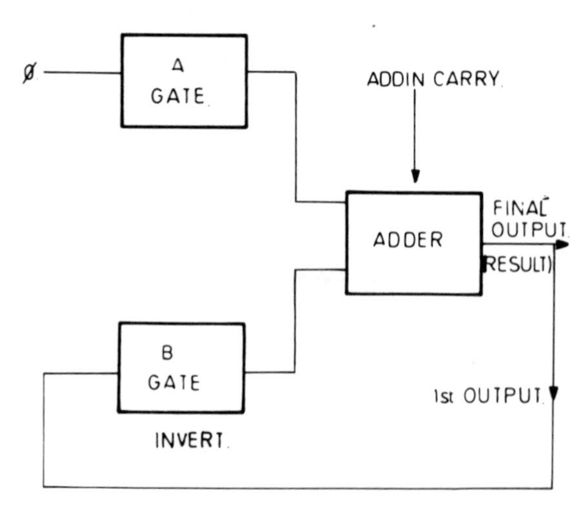

Gray Code

The world is an analog one. In electronics, therefore, digital conversion is often required. One means of converting from linear or rotational movement to digits is by use of a position-sensing code whereby only one bit at a time changes. This has the added advantage of reducing errors as well. The *Gray code* is such a code.

One of the typical applications of Gray code is altitude encoding in secondary-surveillance radar systems.

Decimal	*Gray*
0	0000
1	0001
2	0011
3	0010
4	0110
5	0111
6	0101
7	0100
8	1100
9	1101
10	1111
11	1110
12	1010
13	1011
14	1001
15	1000

Conversion of Decimal to Gray Code (TGC)

Example 4.27

Convert 45_{10} to Gray code.

1. First convert to binary $45_{10} = 00101101_2$. (Note that the leading zero must be included.)
2. To convert to TGC modulo 2 (EXCLUSIVE OR) adjacent bits.

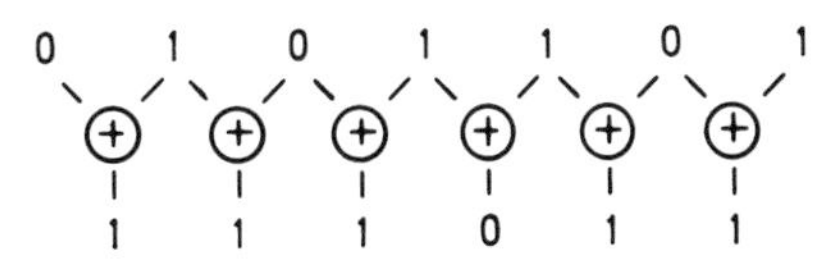

Therefore $45_{10} = 111011_{TGC}$.

Gray to Decimal Conversion

The procedure for this conversion is as follows:

1. Record all bits up to and including first 1.
2. Record 1s up to but not including next 1.
3. Record 0s up to but not including next 1.
4. Record 1s up to but not including next 1, and so on.
5. Convert from binary to decimal.

Example 4.28

Convert 001001011_{TGC} to decimal.

001 (1)

00111 (2)

0011100 (3)

001110010_2 (4)

Convert to decimal:

$$001110010_2 = 114_{10}$$

Exercise 4.8

1. Convert the following to Gray code:
   a. 01011_2
   b. 1011011_2
2. Convert the following to binary:
   a. 10101110_{TGC}
   b. 111011101_{TGC}

Table 4–2
Comparison of Number Systems

Base 16	*Base 10*	*Base 9*	*Base 8*	*Base 7*	*Base 6*	*Base 5*	*Base 4*	*Base 3*	*Base 2*
0	0	0	0	0	0	0	0	0	0
1	1	1	1	1	1	1	1	1	1
2	2	2	2	2	2	2	2	2	10
3	3	3	3	3	3	3	3	10	11
4	4	4	4	4	4	4	10	11	100
5	5	5	5	5	5	10	11	12	101
6	6	6	6	6	10	11	12	20	110
7	7	7	7	10	11	12	13	21	111
8	8	8	10	11	12	13	20	22	1000
9	9	10	11	12	13	14	21	100	1001
A	10	11	12	13	14	20	22	101	1010
B	11	12	13	14	15	21	23	102	1011
C	12	13	14	15	20	22	30	110	1100
D	13	14	15	16	21	23	31	111	1101
E	14	15	16	20	22	24	32	112	1110
F	15	16	17	21	23	30	33	120	1111
10	16	17	20	22	24	31	100	121	10000
11	17	18	21	23	25	32	101	122	10001
12	18	20	22	24	30	33	102	200	10010
13	19	21	23	25	31	34	103	201	10011
14	20	22	24	26	32	40	110	202	10100
15	21	23	25	30	33	41	111	210	10101
16	22	24	26	31	34	42	112	211	10110
17	23	25	27	32	35	43	113	212	10111
18	24	26	30	33	40	44	120	220	11000
19	25	27	31	34	41	100	121	221	11001
1A	26	28	32	35	42	101	122	222	11010
1B	27	30	33	36	43	102	123	1000	11011
1C	28	31	34	40	44	103	130	1001	11100
1D	29	32	35	41	45	104	131	1002	11101
1E	30	33	36	42	50	110	132	1010	11110
1F	31	34	37	43	51	111	133	1011	11111
20	32	35	40	44	52	112	200	1012	100000

5 Bistables and Monostables

Bistables

A bistable or *bistable multivibrator,* as it is more properly called, is a sequential device. Its outputs may, at any instant, be in one of two stable states, hence the name *bi-* (meaning "two") *stable.* Alternative names for bistables are flip-flops or latches. They are used in a number of applications, including medium-term memory, delaying signals, pulse counters, serial-to-parallel converters, and parallel-to-serial converters.

Consider the resistor-transistor circuit for the NOT gate given in figure 1-11. If two of these are cross-coupled, as demonstrated by figure 5-1, then they form a bistable multivibrator.

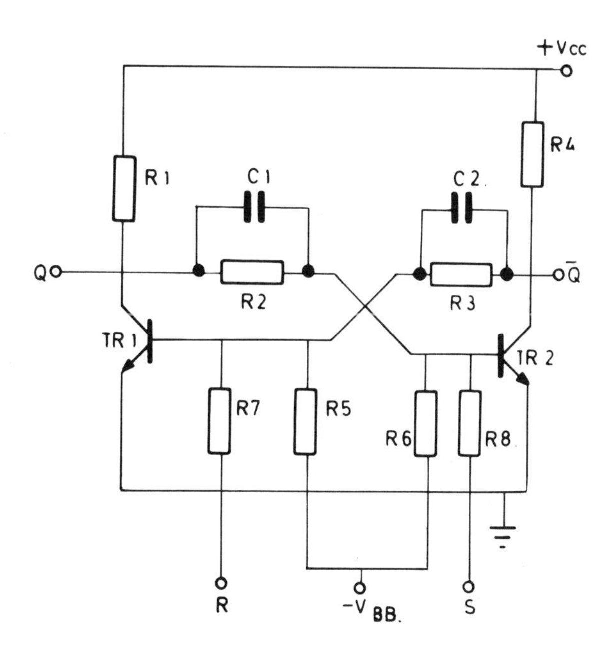

Figure 5-1. The Electronic Circuit of a Bistable Multivibrator

Bistable-Circuit Operation

Assume that TR1 is saturated (on); then Q will be low. If Q is low, then the base of TR2 is held negative by the potential divider R_2 and R_8, and therefore TR2 is off.

Thus $\overline{Q}$ will be high, and so will the base of TR1 holding it on. The circuit remains in this state until a positive pulse is applied to S to change its state. TR2 turns on, $\overline{Q}$ goes low forcing the base of TR1 low, which in turn switches it off. Q is now high and $\overline{Q}$ low. If the positive signal applied to S is now removed, then the circuit remains in this state until a positive signal is applied at R to reverse the state.

This type of bistable is actually called an RS bistable because the application of a high to the R input *resets* the Q output to a low or 0, and the application of a high to the S input *sets* the Q output to a high or a 1.

The RS Bistable

In a digital circuit an RS bistable would not actually be constructed in the manner shown in figure 5-1, but its principle of operation is the same. However, it may be constructed by crosscoupling two NAND gates or two NOR gates, as shown in figure 5-2.

As for the resistor-transistor circuit, there are two inputs R and S and also two outputs Q and $\overline{Q}$. The convention for labeling the outputs of bistables are Q and $\overline{Q}$, which suggests that one is always the inverse of the other. Care must be taken, however, since although this it the general rule it is not always the case.

Before analyzing the circuit function it must be remembered that a NOR gate's output is a 1 if both inputs are 0 and it is a 0 for a 1 on either or both inputs.

Consider that input R is 0 and S is 1. $\overline{Q}$ must be 0, and hence the top NOR gate will have two 0s on its inputs, causing the output Q to be a 1 ($R = 0$, $S = 1$, then $Q = 1$, $\overline{Q} = 0$).

If S changes to a 0 while R remains at 0, then R has not changed, so Q will remain at 1. Hence the bottom gate no longer has a 1 on one of its inputs, although the other is 0, and its output $\overline{Q}$ will therefore be 0. This 0 is crosscoupled to the top gate's input, which means that it now has two 0s on its inputs, producing a 1 output. Thus Q is a 1 and $\overline{Q}$ is a 0 if S changes from a 1 to a 0 and R remains at 0, that is, the outputs remain in their previously switched states ($R = 0$, $S = 0$, then outputs remain the same).

If this procedure is carried out for $R = 1$ and $S = 0$ initially, and for R changing to a 0, then it will also be found that this statement is true.

Consider now that R goes to a 1 and S remains at a 0. For this case R

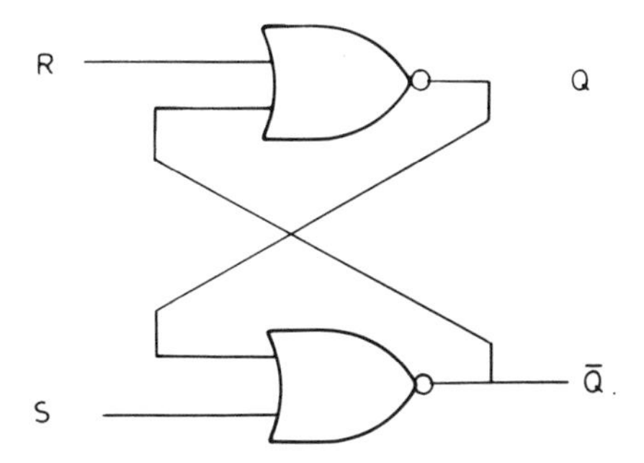

Figure 5-2. RS Bistable Constructed from Two NOR Gates

has changed and presented a 1 to the top NOR gates input, producing a 0 at the Q output. This is cross-coupled to one of the bottom gate's inputs. Both inputs of this gate are therefore now 0, and hence $\overline{Q}$ will be a 1. To check that this is a stable state, the inputs to the top gate must be considered again; they are now both 1, which means that Q will in fact be a 0, as previously stated ($R = 1, S = 0$, then $Q = 0$ and $\overline{Q} = 1$).

If S now becomes a 1 and R remains at a 1, then both NOR gates have a 1 on at least one of their inputs and Q and $\overline{Q}$ will be 0 ($R = 1, S = 1$, then $Q = 0, \overline{Q} = 0$).

Finally, if R changes back to a 0 and S remains a logic 1, then this is the first case considered and hence Q will be a 1 and $\overline{Q}$ will be a 0.

From the description given, a truth table may be constructed as follows:

R	S	Q	$\overline{Q}$
0	0	*	*
0	1	1	0
1	0	0	1
1	1	0	0

* = Previously switched state.

If $Q = 1$, then it is said to be *set;* if it is 0, it is said to be *reset.*

Therefore a general rule for the RS bistable is that "applying a 1 to the 'set' input, *sets* the Q output to a 1, and applying a 1 to the 'reset' input *resets* the Q output to a 0," except, of course, when a 1 is applied to both the "set" and the "reset" inputs at the same time.

It was mentioned previously that the RS bistable can also be constructed from NAND gates. This is shown in figure 5-3. However, the truth table for the NAND gate is different from that of the NOR gate, and this in turn will affect the truth table for the RS bistable.

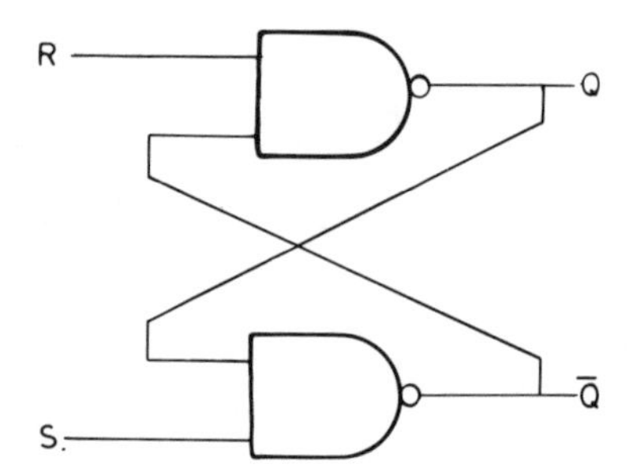

Figure 5-3. RS Bistable Constructed from NAND Gates

The new truth table is:

R	S	Q	$\overline{Q}$
0	0	1	1
0	1	1	0
1	0	0	1
1	1	*	*

* = Previously switched state.

Basically the operation of the circuit is identical except that two 0s on the input produce a two-1s output, instead of causing the output to remain in its previously switched state; and it is now a two-1s input that produces the no-change state.

An Application for the RS Bistable

As mentioned previously, bistable elements can be used as a short-term memory because once their outputs are in one of their stable states they will remain in that state until altered by a change at the inputs. An RS bistable can also be used to remove the "contact bounce" that is generated when a switch is changed over. Although this may not be a problem in many electronic circuits, it can be disastrous for logic circuits; and its removal is therefore important. Many logic circuits change their output states on detecting a "leading" or "trailing" edge (that is, a positive- or negative-going edge) of a pulse. Thus if "bounce" is present, the output may change its state at an incorrect time, leading to erroneous outputs. A binary counter is an example of such a device that changes state on a leading or trailing edge of a pulse input. It will, in fact, increment for each edge detected. Consequently, if contact bounce is present on this input, it may well increment twice or more for one pulse input, thereby generating an incorrect count.

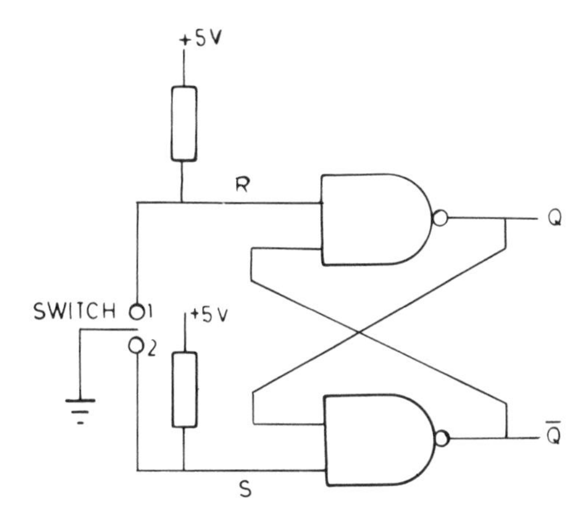

Figure 5–4. Circuit to Remove Contact Bounce

By connecting an RS bistable onto the switch, as shown in figure 5–4, this contact bounce can be removed.

The switch is used to connect either the R input (when in position 1) or the S input (when in position 2) to earth, that is, a logic 0. Whichever input is not connected to ground is at a logic 1 by virtue of its respective pull-up resistor R1 or R2.

With the switch in position 1, Q will be a 1, and with it in position 2, Q will be a 0. Any bounce on changeover will be in a positive direction; since the other input has a logic 1 on it, the bistable will effectively have two 1s on its input (at the instant bounce occurs). This is the condition for no change in the output. Therefore, the bounce will cause no change at Q and consequently will not be present at the input to the following circuitry.

Clocked Flip-Flops

Consider the circuit of figure 5–5. It is actually an RS bistable with additional gating on its inputs.

If the clock input is low, the inputs A and B will not affect the RS inputs. These will both be permanently high, holding Q and $\overline{Q}$ in their previously switched state. However, if the clock is high, then gates 1 and 2 are enabled and inputs A and B will affect the output. Consequently, the outputs only change state when a clock pulse is present, and it is therefore said to be a clocked flip-flop or bistable.

Assume that inputs A and B are only changed when the clock is low, never during the period when it is high. For the purposes of this example, also let t_n = time before the clock pulse and t_{n+1} be the time after the clock leading edge. A and B will actually affect the output as soon as the clock goes high, and it can therefore be said that this is an "edge-triggered" flip-flop.

A truth table can now be drawn to show the action of this circuit, as follows:

t_n		t_{n+1}	
A	B	Q	
0	0	Q_n	(The Q output will remain as it was before the clock pulse, that is, at time t_n.)
0	1	0	(Q goes to a 0 and remains at a 0 until both a change in A or B and another clock pulse occur.)
1	0	1	(Q goes to a 1 and remains at a 1 until both a change in A or B and another clock pulse occur.)
1	1	?	(The output is indeterminate. In this case R and S are both 1 after the clock pulse, and hence the output is indeterminate.)

The D-Type Flip-Flop

The D-type flip-flop is also a clocked flip-flop. Its circuit symbol is shown in figure 5–6.

As can be seen from the circuit symbol, the D-type has four inputs (data in, clock, preset, and clear) and two outputs (Q and $\overline{Q}$). The data input is sometimes referred to as the *steered input,* as it causes a change only when a clock is applied to the device. The preset and clear are referred to as the *forced inputs,* as they actually "force" the output into a particular state, regardless of what is present at the data and clock inputs.

If preset or clear are at a 0, then the D-type's operation is the same as that for the RS bistable constructed from NAND gates, with the RS inputs being replaced by preset and clear. When preset and clear are both 0, then the Q and $\overline{Q}$ are both 1. If preset is 0 and clear is 1, then the Q output is *preset* to a 1 and $\overline{Q}$ will be its inverse, 0. If clear is a 0 and preset a 1, then the Q output is *cleared* to a 0 and $\overline{Q}$ is a 1. In the final possible case, if preset and clear are both 1, then the outputs will remain unchanged until a positive-going clock edge is received at the D-type's clock input. At this time the data on the D input is clocked through to the Q output, and $\overline{Q}$ becomes its inverse. Therefore, a 0 on the preset input or a 0 on the clear input will override the clock, and any change in either the clock or the data input will not affect the output. It is only when preset and clear are both at 1 that the data input affects the output.

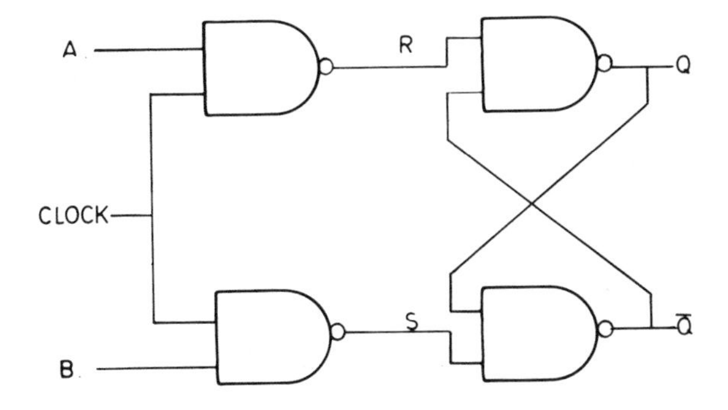

Figure 5-5. A Clocked Flip-Flop

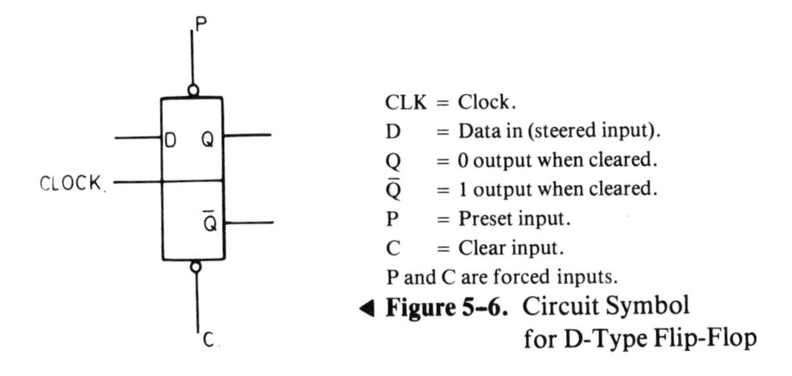

CLK = Clock.
D = Data in (steered input).
Q = 0 output when cleared.
$\bar{Q}$ = 1 output when cleared.
P = Preset input.
C = Clear input.
P and C are forced inputs.

◀ **Figure 5-6.** Circuit Symbol for D-Type Flip-Flop

The overall functioning of the D-type is shown by the following truth table:

		t_n	t_{n+1}	
P	C	D	Q	$\bar{Q}$
0	0	X	1	1
0	1	X	1	0
1	0	X	0	1
1	1	1	1	0
1	1	0	0	1

where t_n = Bit time before the clock pulse.

t_{n+1} = Bit time after the clock pulse.

It is important to note that the data transferred to the Q output is the data that was present at the D input *just before the clock edge.* Therefore, should the data change at exactly the same time as a clock transition from 0 to 1 (that is, the positive-going edge of the pulse) the D-type will not clock through whatever the data change was. In effect, this change will not be "seen" by it. This is demonstrated in figure 5-7, which is a timing diagram for the D-type's operation. This diagram gives examples of both where the data is stable at the clock-edge time and where there is a transition. Note also from this diagram that the output only changes when there is a clock leading edge and that after the clock edge it remains in its previously switched state until another clock edge and a change in the data occurs, or alternatively until preset or clear go low.

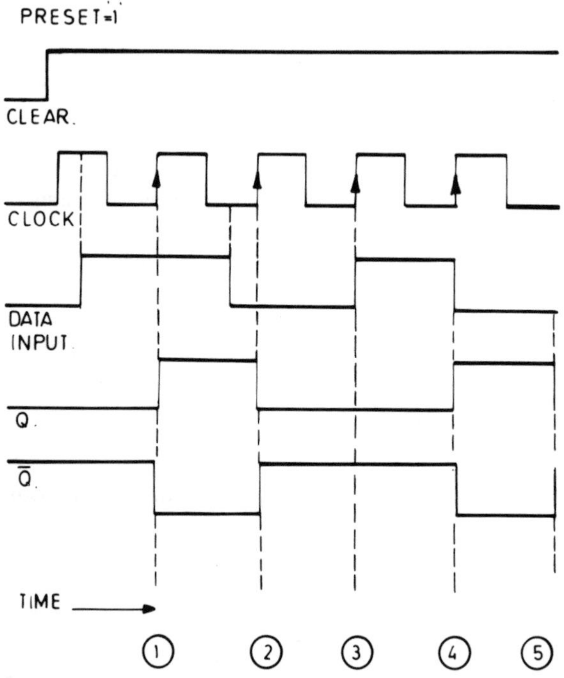

(1) Data just before clock edge = 1; therefore Q = 1 after clock-edge time.
(2) Data just before clock edge = 0; therefore Q = 0 after clock-edge time.
(3) Data transition coincident with clock edge but data was a 0 *just before the clock edge* and therefore it is a 0 that is clocked through to the Q output, that is, the output remains the same in this case.
(4) Once again the data transition is coincident with the clock edge. However the data input was a 1 *just before the clock edge* and therefore a 1 is clocked through to the Q output.
(5) Data input a 0 just before the clock edge and therefore Q = 0 after the clock edge.

Figure 5-7. Timing Diagram for D-Types Basic Clocked Operation

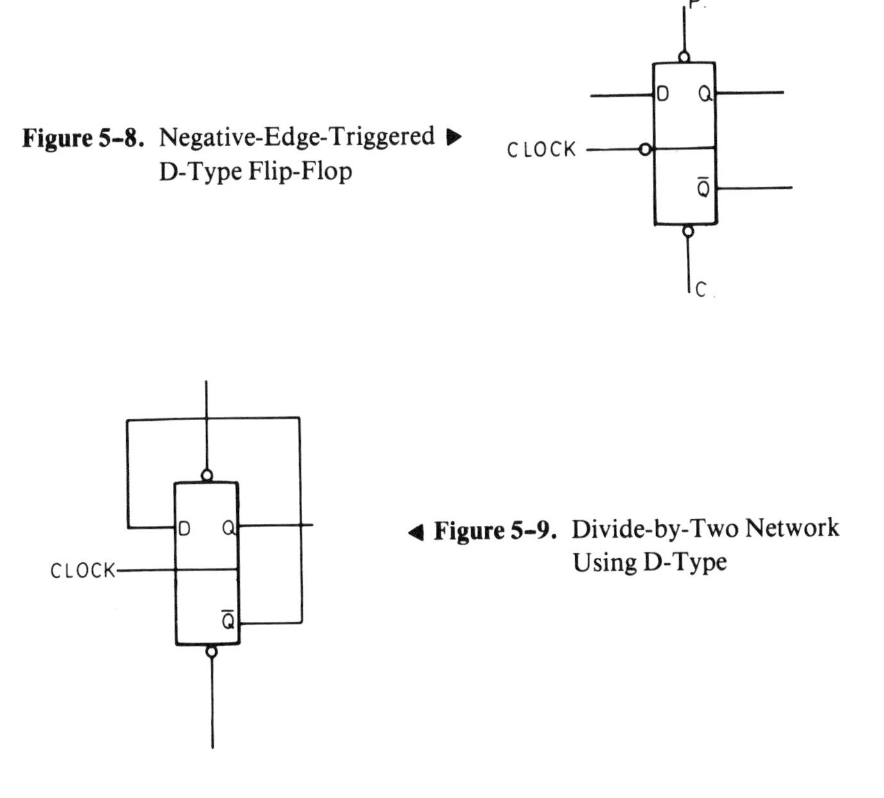

Figure 5-8. Negative-Edge-Triggered D-Type Flip-Flop

Figure 5-9. Divide-by-Two Network Using D-Type

So far it has been stated that the D-type is clocked on the positive-going edge, and this is indeed the case for the circuit symbol shown. However, it is possible to obtain a D-type that triggers on the negative-going or falling edge of the clock pulse. The circuit symbol for it is given in figure 5-8. It is identical to that for the positive-edge-triggered flip-flop except that it has a small circle on its clock input, signifying that it is the inverse.

By connecting a D-type in the manner shown in figure 5-9, a divide-by-two circuit is obtained, the output Q being half the frequency of the clock input.

Because the $\overline{Q}$ output is fed back to the D input, the D-input is changing continuously from a 0 to a 1 or a 1 to a 0 on each clock, which means that Q is also changing. This is demonstrated by the timing diagram shown in figure 5-10.

As previously mentioned, a number of D-type flip-flops may be connected together to form a counter, to provide data storage, or as a delay network. Exactly how this is done will be shown in chapter 6.

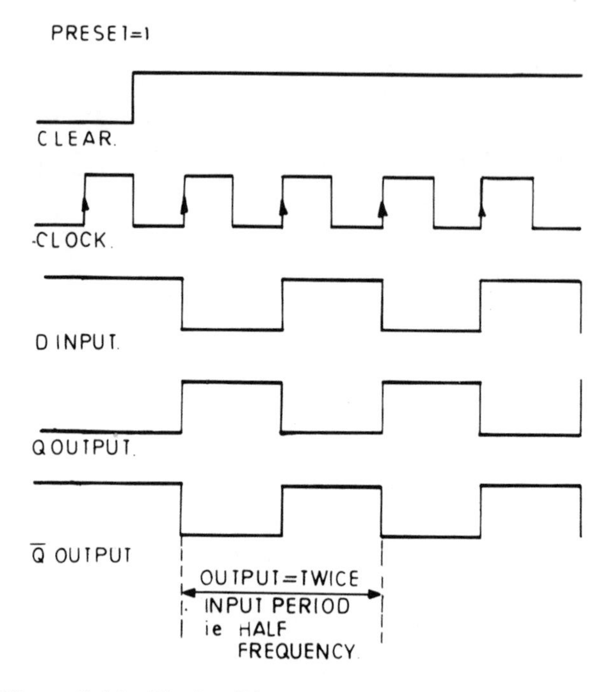

Figure 5–10. Timing Diagram for Divide-by-Two Circuit

The JK Bistable

Two types of JK bistable exist, the *edge-triggered* JK and the *master-slave* JK. The basic circuit symbol for both types is the same, as shown in figure 5–11. Because of differences in their operation, however, they will be discussed separately in this section.

The operation of preset and clear inputs on the JK is identical to that for the D-type, with the same requirement that they must both be a 1 for clocked operation.

The Edge-Triggered JK Bistable

Consider the circuit for the basic clocked RS bistable in figure 5–5. Although it functioned adequately, its limitation was that when inputs A and B were both high, the output was indeterminate. If this circuit is modified to the form shown in figure 5–12, by introducing feedback paths from Q and $\overline{Q}$ outputs and relabeling the inputs J and K, then the problem is

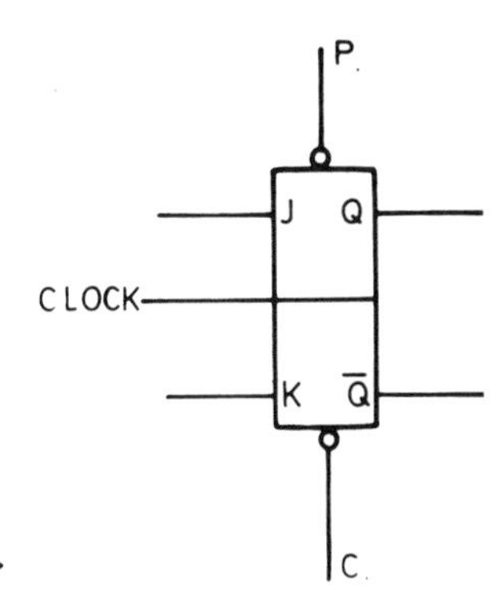

CLK = Clock input.
J and K = Steered inputs.
P = Preset input.
C = Clear input. P and C are forced inputs.

Figure 5-11. Circuit Symbol for JK Bistable ▶

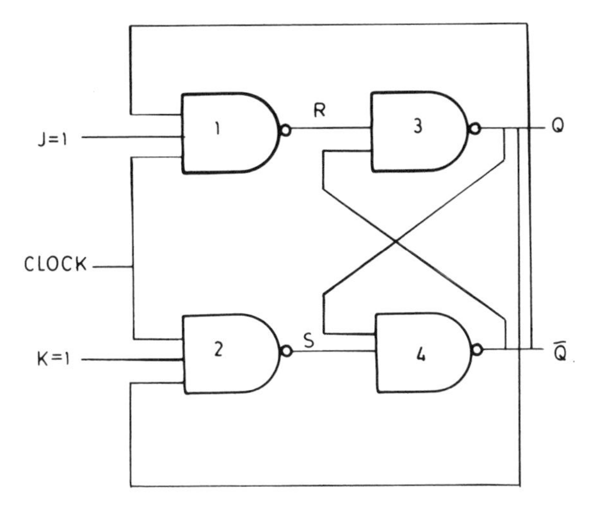

Figure 5-12. Simple Edge-Triggered JK

overcome and the circuit is that of a simple edge-triggered JK without preset and clear inputs.

The operation of the circuit may be considered as follows: Assume that the last states of Q and $\overline{Q}$ were 0 and 1 respectively and that J = K = 1. Then when a clock pulse is applied to the circuit, NAND-gate (1) will have three 1s on its input and will output a low. NAND-gate (2) will output a high. These outputs are presented to the R and S inputs of the following RS bistable constructed from gates (3) and (4). This causes the outputs to "toggle," that is, Q changes from a 0 to a 1 and $\overline{Q}$ from a 1 to a 0.

Previously, of course, because no feedback paths were present, R and S both went to a 1 under these input conditions, giving indeterminate outputs.

The actual change in output occurs at the leading edge of the clock, that is, as soon as the clock goes high, and the JK is therefore defined as *edge-triggered.* If further clock pulses were applied to this circuit, the outputs would toggle at each leading edge, causing the JK to act as a divide-by-two circuit because the Q output would be half the frequency of the clock input.

Thus the overall truth table for the edge-triggered JK bistable is:

			t_n		t_{n+1}	
P	C	J	K	Q_n	Q_{n+1}	$\overline{Q}_{n+1}$
0	0				1	1
0	1				1	0
1	0				0	1
1	1	0	0	Q_n	Q_n	$\overline{Q}_n$
1	1	0	1	X	0	1
1	1	1	0	X	1	0
1	1	1	1	Q_n	$\overline{Q}_n$	Q_n

where t_n = Bit time before clock leading edge.

t_{n+1} = Bit time after clock leading edge.

X = Don't care.

It can be seen from the truth table that when J and K are both zero and a clock is applied, the output remains in the same state as before the clock edge. However, if J = 0 and K = 1 or J = 1 and K = 0, then the clock effectively causes J to be clocked through to Q and K to be clocked through to $\overline{Q}$. A timing diagram for the JK's operation is shown in figure 5-13.

A JK may alternatively be negative-edge triggered. In this case a small circle will be drawn on the clock input.

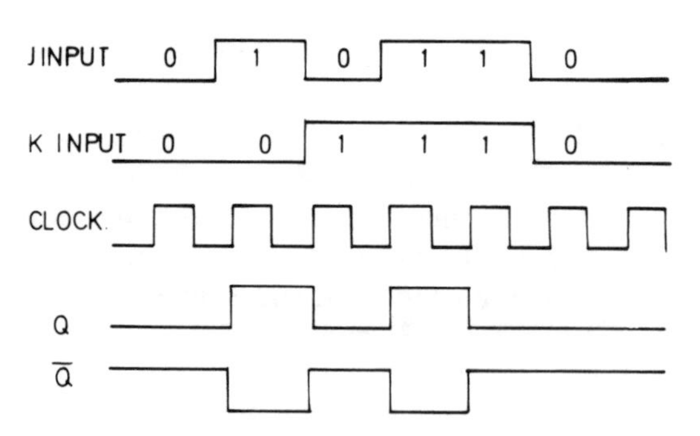

Figure 5-13. Timing Diagram

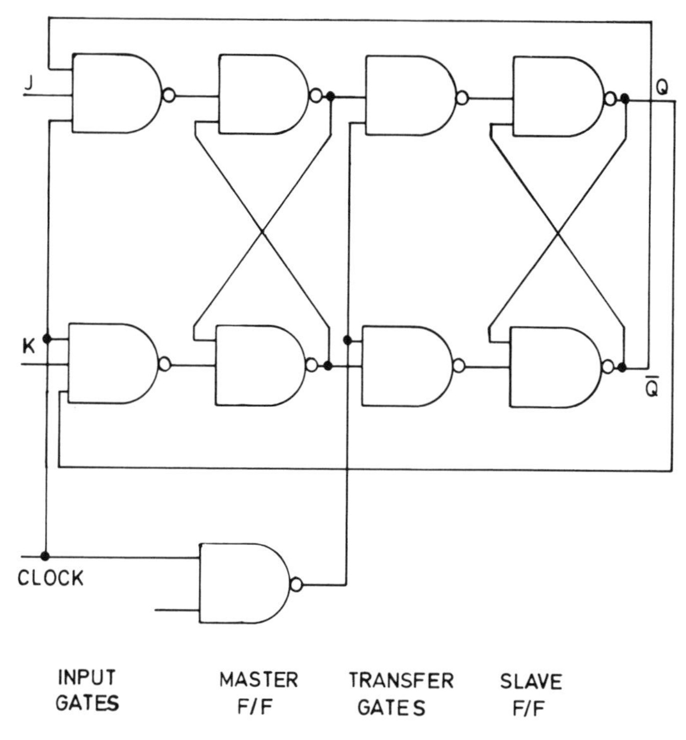

Figure 5–14. A Master-Slave JK

The Master-Slave JK Bistable

The basic operation of the master-slave JK is the same as that for the edge-triggered except that the changes in output occur because of a clock pulse input as opposed to just a leading or trailing edge.

The device itself can be divided up into two parts, the "master" and the "slave," as shown in figure 5–14.

The circuit's operation is as follows: When the clock goes high, the second or slave flip-flop is disabled by the inverted clock input, but the first or master flip-flop is set or reset depending on the J and K inputs. When the clock goes low, the master flip-flop is disabled but the slave is enabled and responds to its inputs, that is, the outputs of the master flip-flop. This means that data presented to the JK inputs while the clock input is low will not affect the output until a complete clock pulse has been supplied to the JK. Note that this input must be maintained until the clock goes high. The timing diagram shown in figure 5–15 demonstrates this operation.

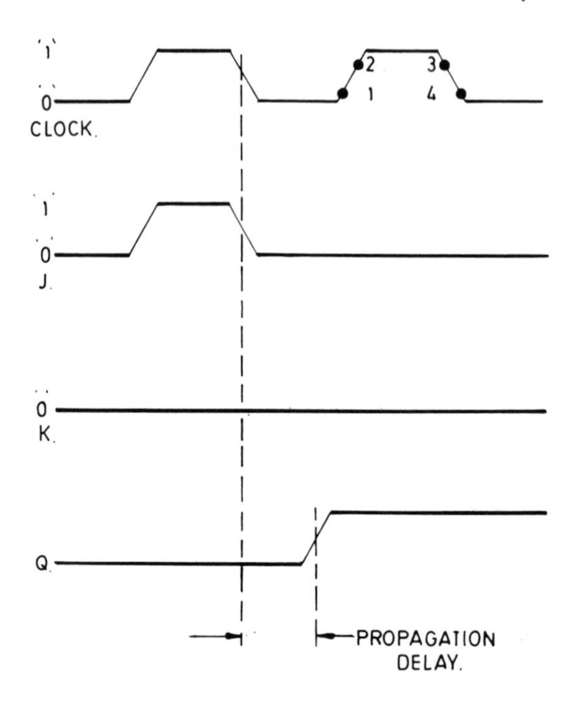

Figure 5-15. Timing Diagram for Master-Slave JK

The overall operation can be summarized by the actions that occur at the points numbered on the timing diagram, as follows:

1. The slave is isolated from the master.
2. Information is entered to the master.
3. The input gates are disabled.
4. Information presented from master to slave and hence output.

JK Bistable with AND-Gate Inputs

Other than the basic JKs described so far in this chapter, there are also JK bistables with AND-gated inputs, as shown in figure 5-16(a). Their operation is the same except that J is actually equal to J1, J2, J3, and J4 all ANDed together; in other words, it will only be high if *all* the J inputs are high. Similarly, the K input will only be high if K1, K2, K3, and K4 are all high. Figure 5-16(b) shows an alternative symbol for the device.

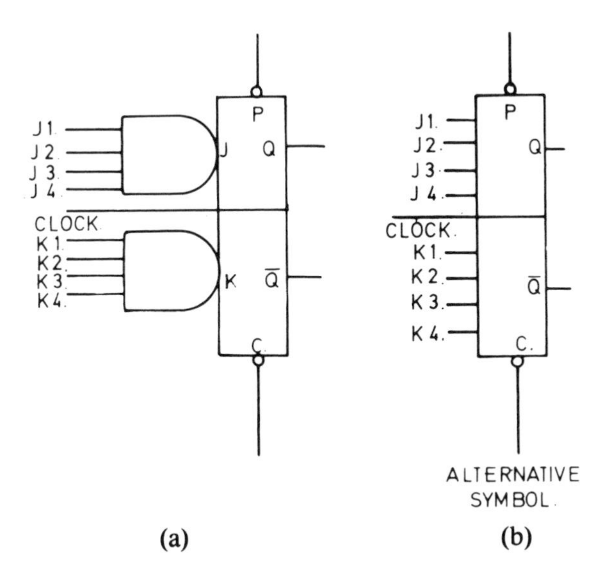

(a) (b)

Figure 5–16. AND-Gated JK Bistable

Monostables

If one of the bias resistors of the bistable circuit of figure 5–1 is replaced by a capacitor, as shown in figure 5–17, then the circuit action is different from that of the basic flip-flop. When it is triggered by a clock pulse, the circuit changes state but only remains in the new state for a time dependent on the value of the capacitor, C, and its resistor, R. Thus a clock pulse applied to a monostable causes its output to change state and to remain in that state for a set time interval before returning to its original state.

Circuit Operation

Prior to the application of the clock pulse, the monostable is in its stable state and TR1 is held off by V_{BB} applied to its base. The collector of TR1 is therefore high, and this is coupled to TR2 base via the C, holding TR2 on. Thus the capacitor C is charged, and the collector of TR2 is low.

When a positive-going input pulse is applied to the clock input, the voltage $-V_{BB}$ is overcome, raising the base potential of TR1 and turning it on. The TR1 collector falls toward earth potential, applying a negative-going voltage to the base of TR2, which switches it off and causes its collector to go high.

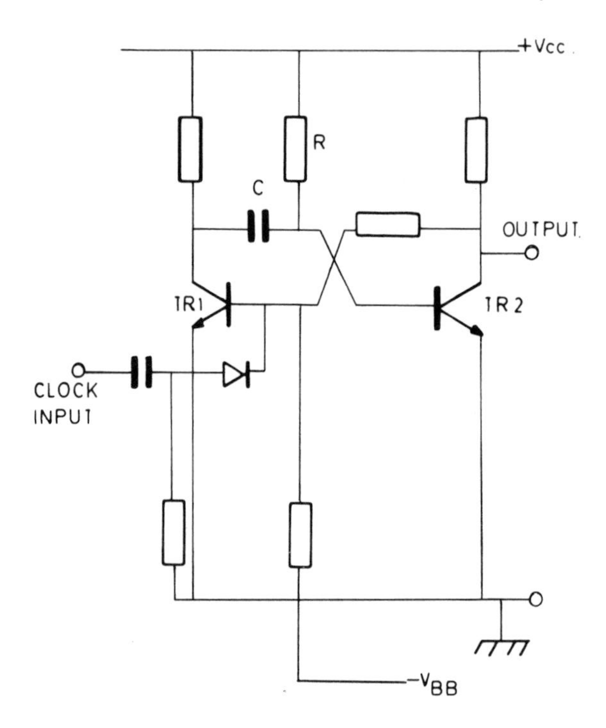

Figure 5-17. Monostable Circuit

TR1 is now effectively a closed switch, which connects the capacitor C to earth and allows it to discharge slowly through R. As the charge across C reduces, the base potential of TR2 becomes more positive until it reaches a sufficiently positive voltage to switch it on once more, which causes the output to go low again. This low is also applied to the base of TR1, switching it off. The capacitor C is now charged to its original level, and the monostable circuit remains in this stable state until the application of another input pulse.

If required, a complementary output can be obtained from the collector of transistor TR1.

A monostable is also called a *one-shot* or a *single-shot.* It can be used to provide output pulses of a duration from 20 nsec to minutes by altering the values of its timing components. Monostables can also be used as pulse shapers. There are two basic types; the *gated* (or *nonretriggerable*] monostable and the *retriggerable* monostable. These are given in the following section, where they are drawn in their logic-symbol form.

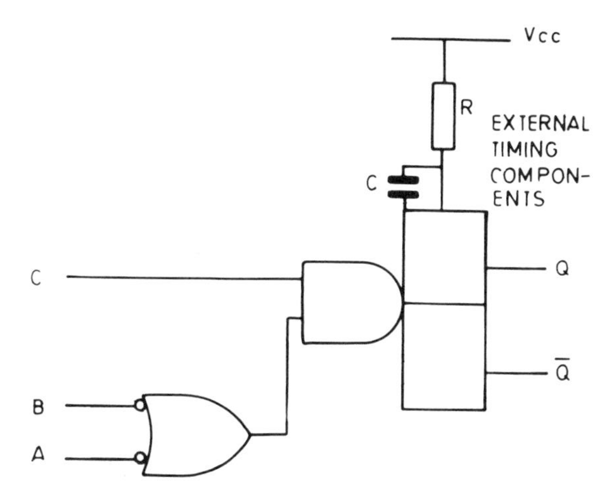

Figure 5-18. A Gated Monostable

The Gated Monostable

As shown in figure 5-18, this monostable will not be triggered unless there is a 1 on both AND-gate inputs. This means that there must be a 1 on C input and a 0 on either A or B inputs. The gate being fed by A and B inputs is sometimes symbolized by a NAND symbol in place of the one shown, but the logic function is the same.

The monostable may be triggered on either the positive-going edge or the negative-going edge of the input waveform, depending on the input to which the waveform is applied. Positive-edge triggering can be obtained by holding either A or B at 0 and applying the waveform to C, as demonstrated in figure 5-19. Negative-edge triggering is obtained by holding C at 1 and applying the waveform to either A or B, as shown in figure 5-20. It must be noted that with this type of monostable there is no retrigger until the end of the time period, which is determined by the external timing components C and R.

The Retriggerable Monostable

This monostable, shown in figure 5-21, is different from the previous one in that applying another pulse input before the first time interval has finished retriggers the device and begins a new time period, as shown in figure 5-22. Thus by continually pulsing its input and selecting timing values to be

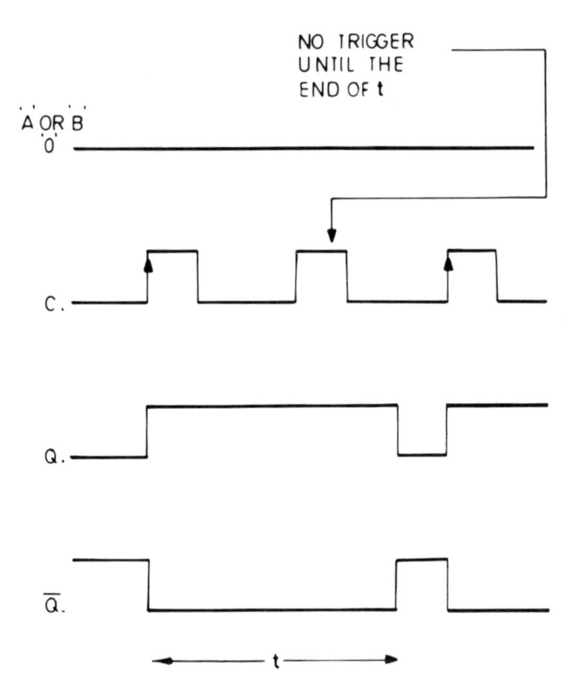

Figure 5-19. Positive-Edge Triggering

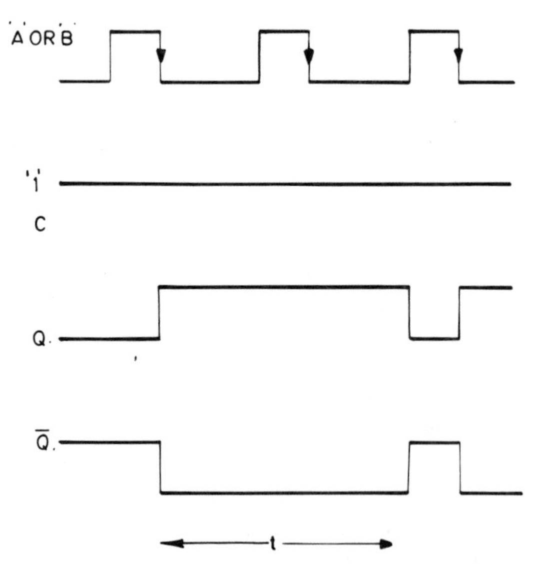

Figure 5-20. Negative-Edge Triggering

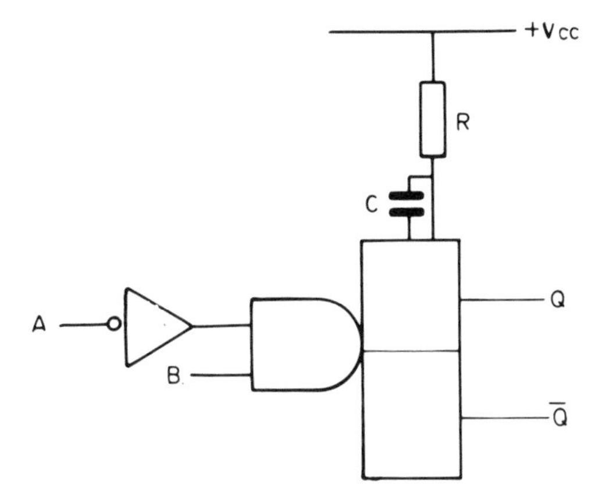

Figure 5-21. Retriggerable Monostable

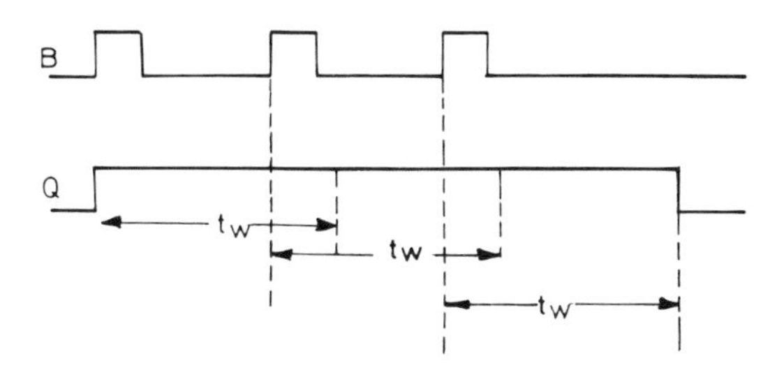

Figure 5-22. Timing Diagram for Retriggerable Monostable

greater than the input pulse period, its output time duration may be lengthened indefinitely. One application for this is in determining whether a pulse has not occurred within a certain time period.

As mentioned previously, the duration of the output pulse is dependent on the external capacitor and resistor values. However, the monostables may also have an internal timing resistor that can be used for short time periods.

Alternatively, for longer time durations diodes or transistors may be connected to the external-timing-component pins of the monostable, as shown in figure 5-23(a) and (b).

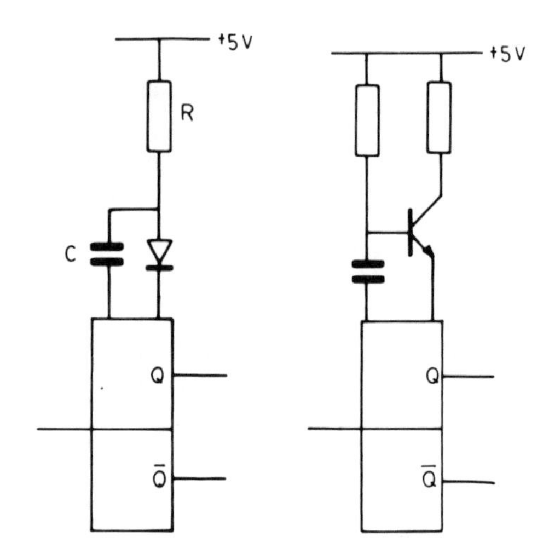

Figure 5-23. Monostables for Longer Time Periods

Exercise 5.1

1. Draw the waveform for the Q output, given R and S inputs.

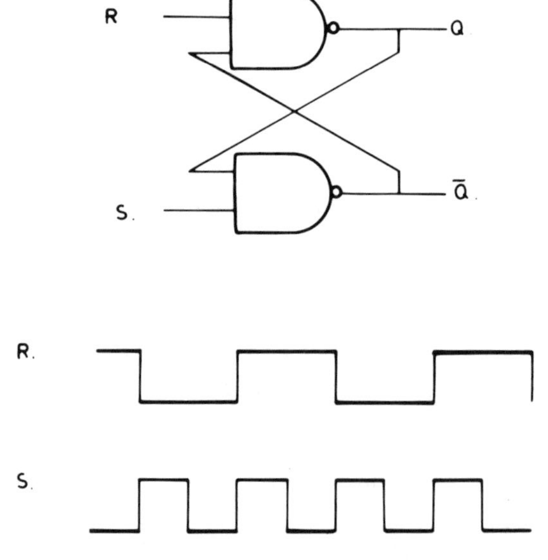

2. Given the following circuit and associated input waveforms, draw the waveform of the Q_2 output. What is the action of this circuit?

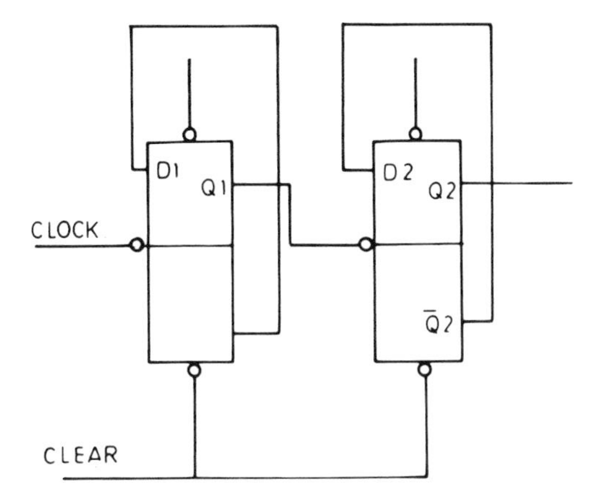

The presets are connected to + 5 v via pull-up resistors.

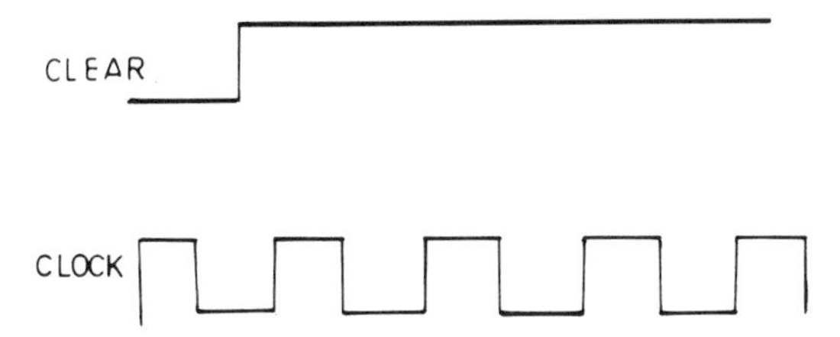

3. Given the following circuit and associated input waveforms, draw the waveform of the Q_2 output and hence deduce what the action of the circuit is.

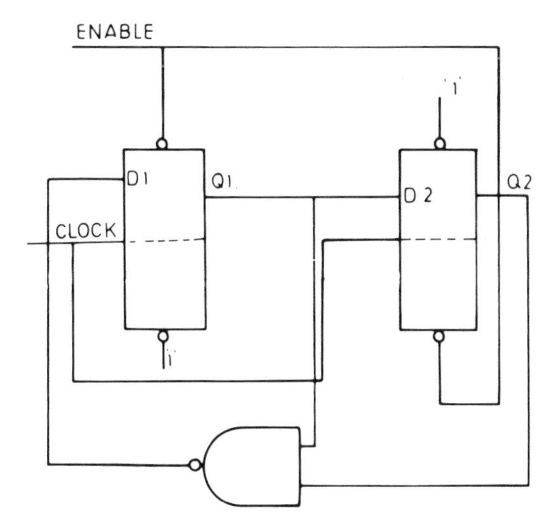

4. Draw the Q output waveform for the following circuit.

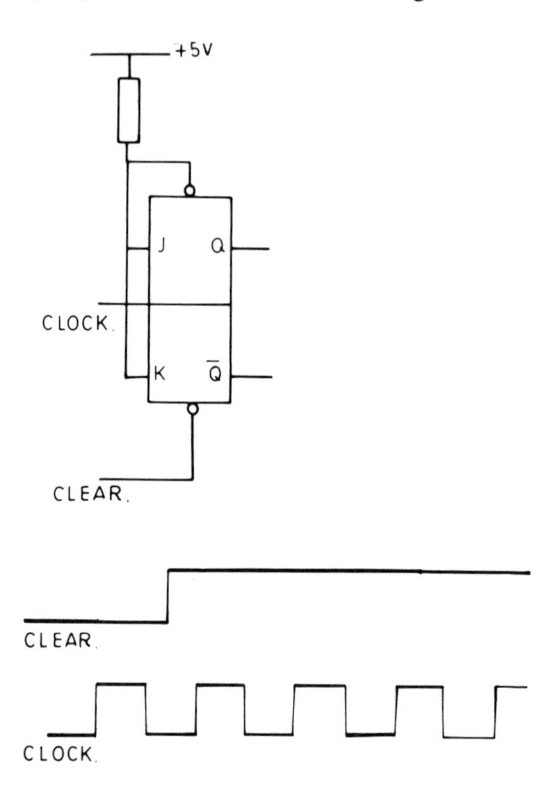

5. Draw the Q-output waveform of the following circuit for a retriggerable monostable.

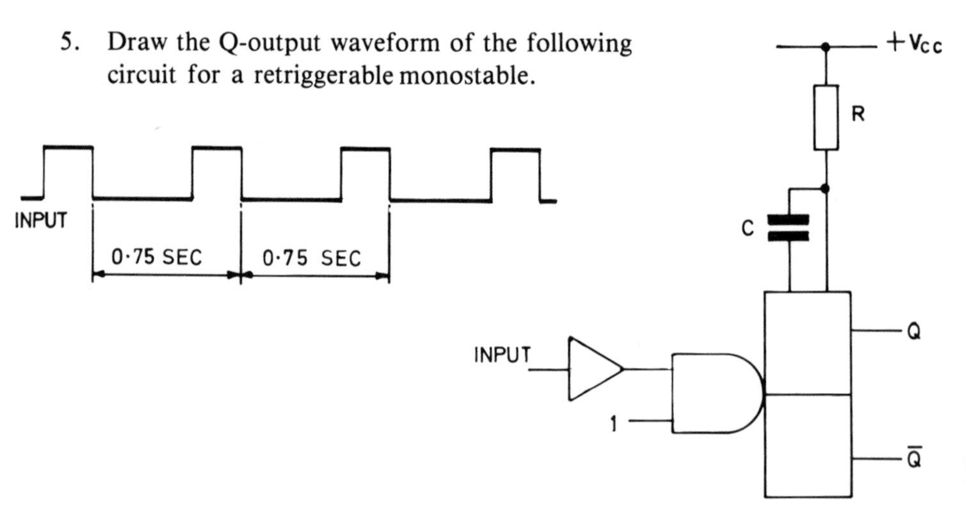

6 Shift Registers and Counters

Shift Registers

A *shift register* consists of a number of flip-flops connected in a manner that allows the shifting of binary data to the left or right one bit at a time.

It is partly because of this basic characteristic and because it is constructed from flip-flops that the shift register has so many applications. For example, it may be used to:

1. Store computer instructions and memory addresses.
2. Store numbers (data to be used for calculations or the results of calculations).
3. Delay information by a specified number of shift commands (clock pulses).
4. Convert a serial stream of bits into a parallel slice of bits.
5. Convert a parallel slice of bits into a serial stream of data.
6. Pulse-train generator.

One of the simplest forms of shift register can be constructed from D-type flip-flops, as shown in figure 6–1. The preset inputs are not shown on this diagram and are therefore assumed to be logic-level 1s.

Consider the 4-bit binary number 1010_2 being applied to the data input in a serial form (least-significant bit first) and a clock waveform applied to the circuit, as demonstrated by the timing diagram, figure 6–2. At the leading edge of each clock pulse the level on the D input, *just before the clock*

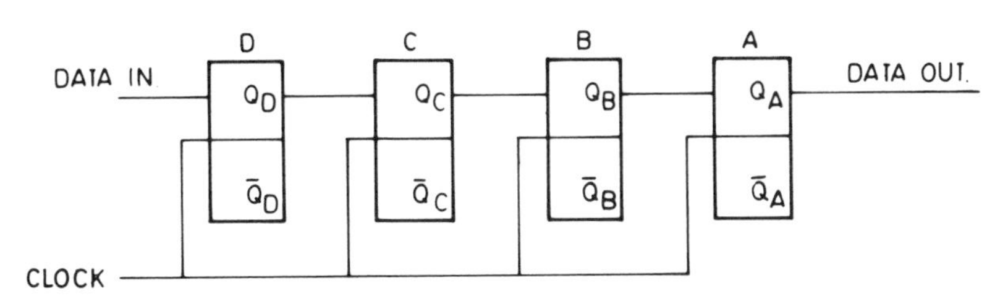

Figure 6–1. Shift Register Constructed from D-Types

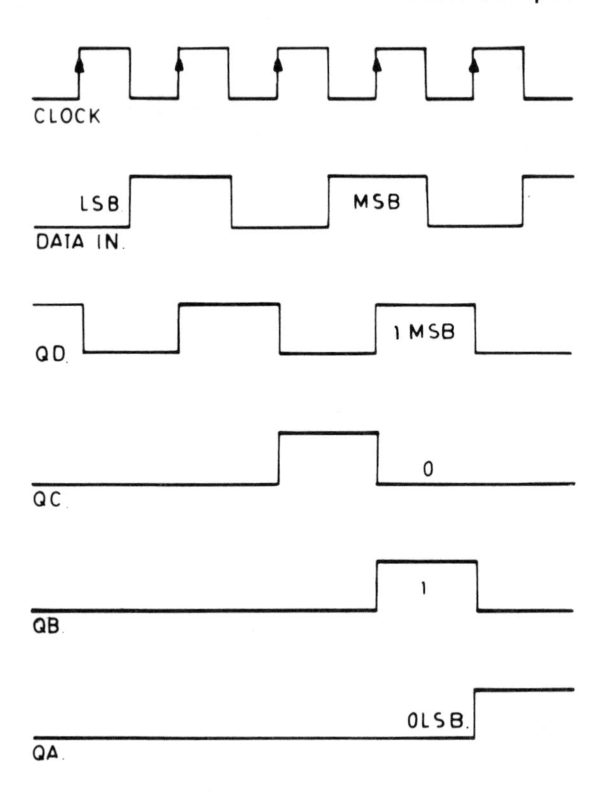

Figure 6–2. Timing Diagram for Shift Register

pulse, is clocked through the Q output. Thus on the first clock edge the least-significant bit (LSB) of the number is clocked through the the Q_D output, and then on the second clock edge it is shifted through to the Q_C output. Hence the LSB is shifted right on each clock edge and so are the other bits of the number. After the fourth clock pulse the complete number is present at the four Q outputs.

If the clock is now stopped, and the power is held on, the flip-flops remain in their respective switched states acting as a store. Also the Q outputs can all be sampled at the same instant, providing the 4-bit number as a parallel output. The shift register has therefore been used as a serial-to-parallel converter.

It may be seen from the timing diagram that the data input has also been delayed by its shifting right down the register; therefore, a shift register may be used to delay data if required. Shift registers may also be constructed using JK bistables, as shown by figure 6–3.

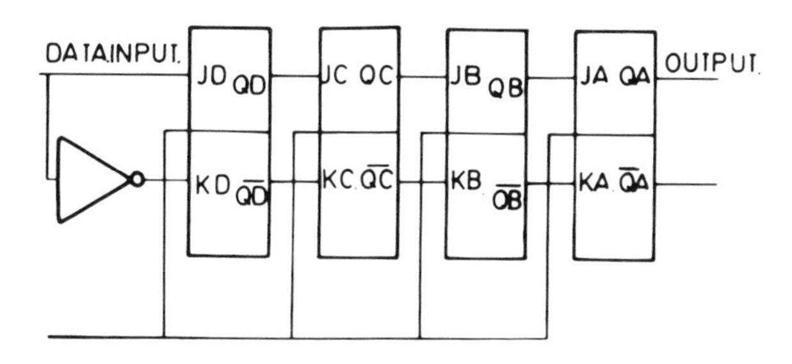

Figure 6–3. A Shift Register Constructed from JK Bistables

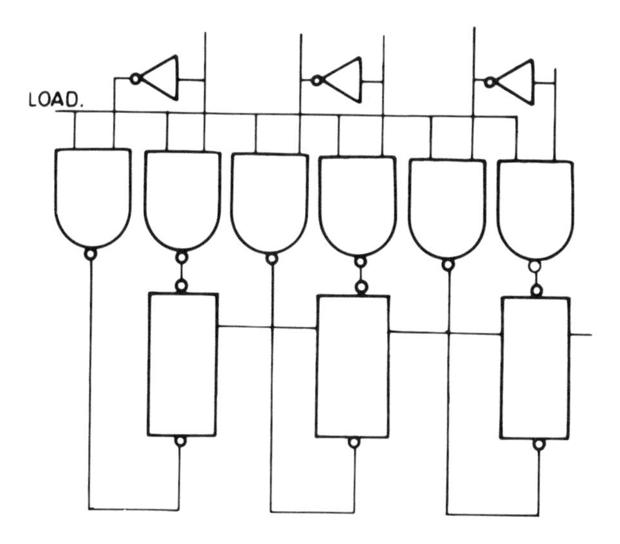

Figure 6–4. Parallel Loading of Shift Register

Parallel-to-serial conversion can be achieved with either type of shift register by "clearing" or "presetting" specific flip-flops within the register and then applying a clock to shift the data out in serial. The loading of the register in parallel can be achieved using gating circuitry such as that shown in figure 6–4.

Alternatively, if storage of parallel data is required, then once the register is loaded no clock need be applied. The data is therefore present at the flip-flop outputs to be sampled when required.

A shift register that can perform all the functions previously described may be shown on a circuit diagram by a symbol such as the one of figure 6–5, which is basically a rectangle with all the inputs and outputs labeled.

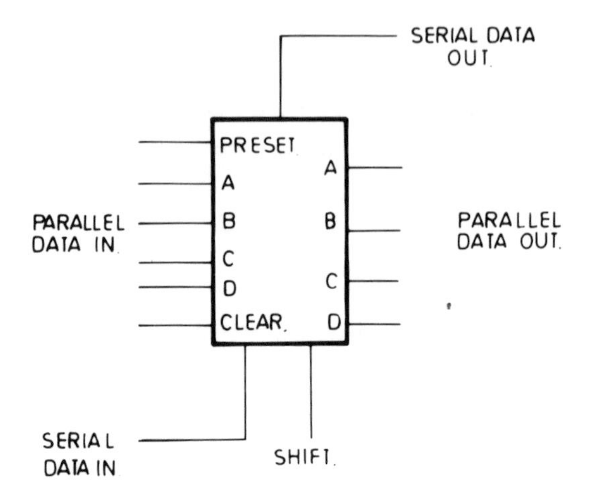

Figure 6-5. General Circuit Symbol for a Shift Register

This shift register has three "control" inputs: *shift, clear,* and *preset.* Their functions are as follows:

Shift: The clock input for the register.

Clear: A logical 1 on the clear input automatically clears the register to 0000 no matter what signals are applied to the other inputs.

Preset: A logical 1 at this input loads the register with the contents of all four data bits, on the parallel in lines, simultaneously. As mentioned previously, this is a useful method of storing data because it is four times faster (for this example) than storing them in a serial fashion. The series storage method requires four shift commands or clock pulses, whereas parallel storage occurs almost instantly and without the need for a series of commands.

A more-advanced shift register is one that will shift data either left or right.

Ring Counters

By connecting the output of a shift register back to its input, the register's contents will be recirculated continuously when supplied with successive clock pulses. In this way they act as a dynamic store for a binary word. They

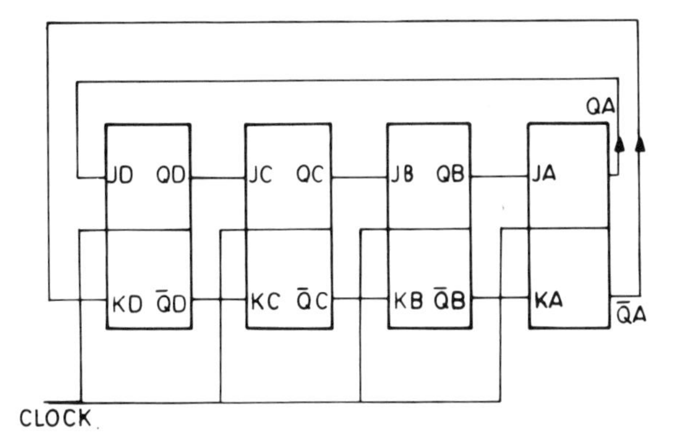

Figure 6–6. Ring Counter

can also act as counters, although nonsequential, because an n-bit counter is capable of providing n possible states. They are then referred to as *ring counters.*

In their most simple form the counters can have a 1 that is continuously circulated. This is shown in figure 6–6, which also gives the states that ensue.

Counters

Counters are used in a wide range of digital devices, including electronic clocks, voltmeters, frequency counters, and, of course, computers.

A counter can be constructed from a number of flip-flops, as with the shift register, but the interconnections between the flip-flops are different. The maximum number of counts is designated by the word *modulo* and is dependent on the number of flip-flops in the counter. For example, a counter constructed from three flip-flops has a modulo of 2^3 or 8 counts, and a counter with four flip-flops a modulo of 2^4 or 16. When the counter reaches its maximum count it is automatically reset to 0 and begins counting up again.

Counters may be designated as either *asynchronous* or *synchronous,* depending on their mode of operation. Asynchronous counters are relatively slow since the output of one flip-flop triggers a change in the state of the next. This is why they are sometimes called *ripple-through* counters. The flip-flops of synchronous counters are clocked by a common clock signal and therefore all receive a trigger pulse at the same time. They are consequently faster than the asynchronous type.

Asynchronous or Ripple-Through Counter

As stated previously, the asynchronous operation is similar to that of a serial shift register in that the status of one flip-flop controls the status of the next. This is the simplest form of counter and may be constructed either from D-type flip-flops, as shown in figure 6-7, or from JKs, as shown in figure 6-8.

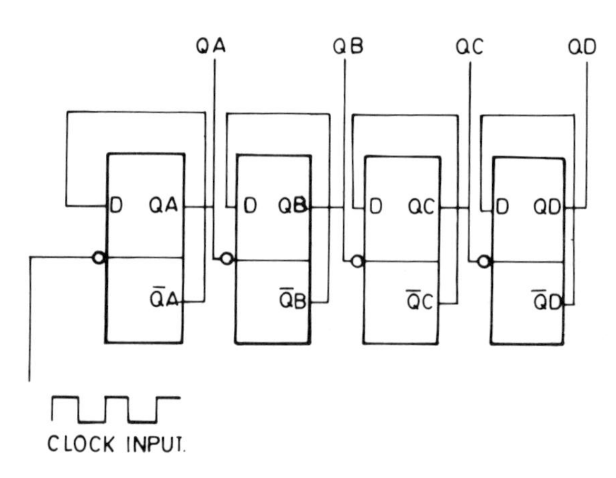

Figure 6-7. Asynchronous Counter Constructed from D-Types

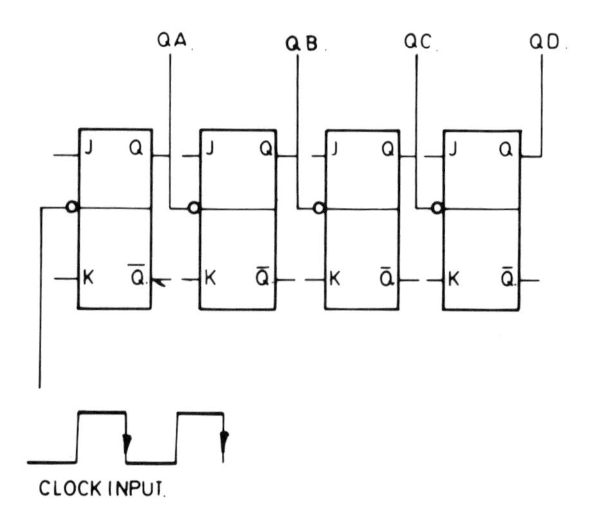

Figure 6-8. Asynchronous Counter Constructed from JK Bistables

Figure 6–7 consists of four D-type flip-flops, each connected as divide-by-two circuits; and figure 6–8 consists of four JKs with their J and K inputs both high, so that each one is a divide-by-two circuit. For both circuits the flip-flops are negative-edge triggered, and each stage is clocked by the Q output of the preceding stage.

A timing diagram for these circuits is given in figure 6–9. Output A toggles for each incoming clock pulse and therefore changes state from a 0 to a 1 every two clock pulses. However, this output is connected to the input of B; thus output B will only change from a 0 to a 1 once every four clock pulses applied to its input, and the outputs Q_D, Q_C, Q_B, and Q_A supply a running total of this count in binary, with Q_A being the least-significant bit.

As may be seen from the timing diagram, the binary number increases with each new clock period.

It should be noted that the modulo-16 counter, as shown, actually counts to 15 decimal, not 16. The reason for this is that 0000 takes one of the count positions, so that although a modulo-16 counter has a full 16 counts, it can in fact only count up to 15.

If the counter of figure 6–8 is now modified slightly in such a way that the outputs are taken not from the JK's Q outputs but from its $\overline{Q}$ outputs, then the count sequence will be down instead of up. Alternatively, the Q outputs can still be sampled but the following stages clocked from the $\overline{Q}$ outputs instead of the Q outputs; then a count-down sequence once again occurs.

The main advantages of asynchronous counters are that they are relatively simple and cheap in comparison with synchronous counters. Their disadvantages are that they are slow and give "false outputs" due to the propagation delay of the gates. This is demonstrated by the timing diagram in figure 6–10. It can be seen from this that for the propagation delay time tP after the second clock input, the Q outputs are actually both 0 for a short period of time instead of changing directly from 01 to 10. For many applications this is not a problem. Where it is, a synchronous counter can be used instead since it does not exhibit the same characteristics.

Synchronous Counters

Synchronous counters use a common clock to all the flip-flops within them, so that each one is triggered at exactly the same instant. Gating is then employed between the outputs of the first stages and the inputs of the latter ones to ensure that a binary count sequence, or any other desired sequence, occurs. Thus they are slightly more complicated and obviously more expen-

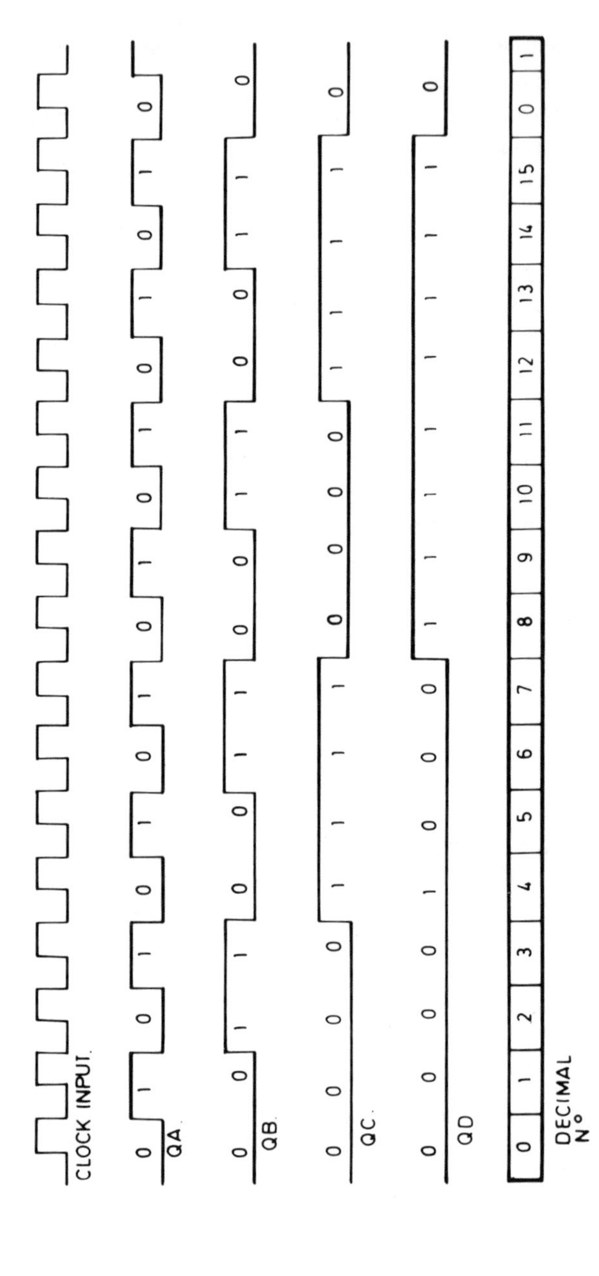

Figure 6-9. Timing Diagram for a 4-Bit Counter

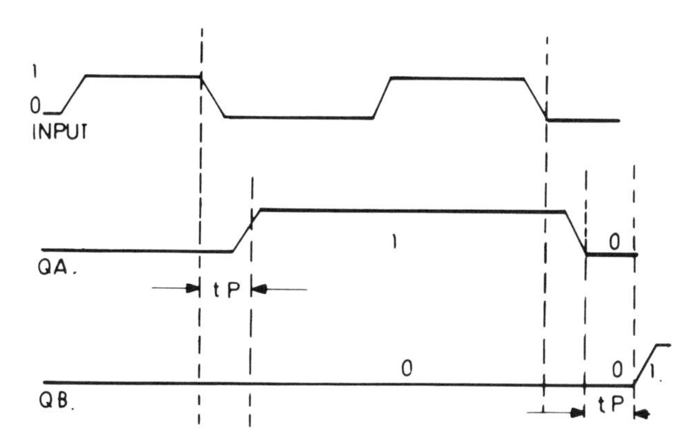

Figure 6–10. Timing Diagram for an Asynchronous Counter Showing Propagation-Delay Effects

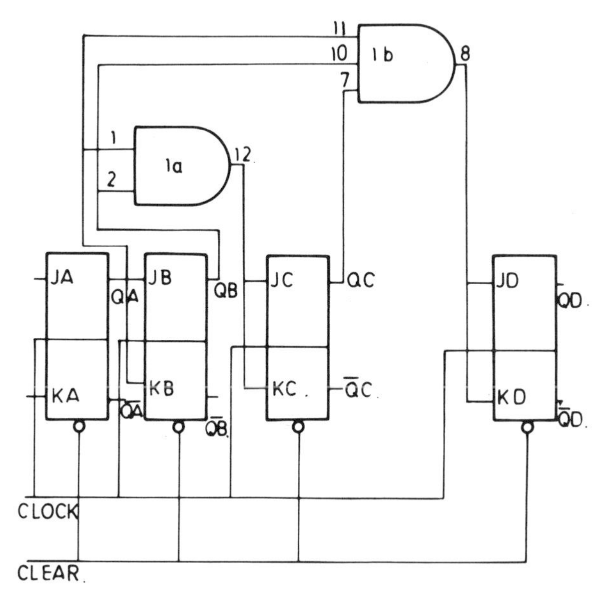

Figure 6–11. Synchronous Modulo-16 Counter

sive than asynchronous counters, but they have the advantage of being faster and of not giving false outputs. An example of a modulo-16 synchronous counter is given in figure 6–11 and its associated timing diagram in figure 6–12.

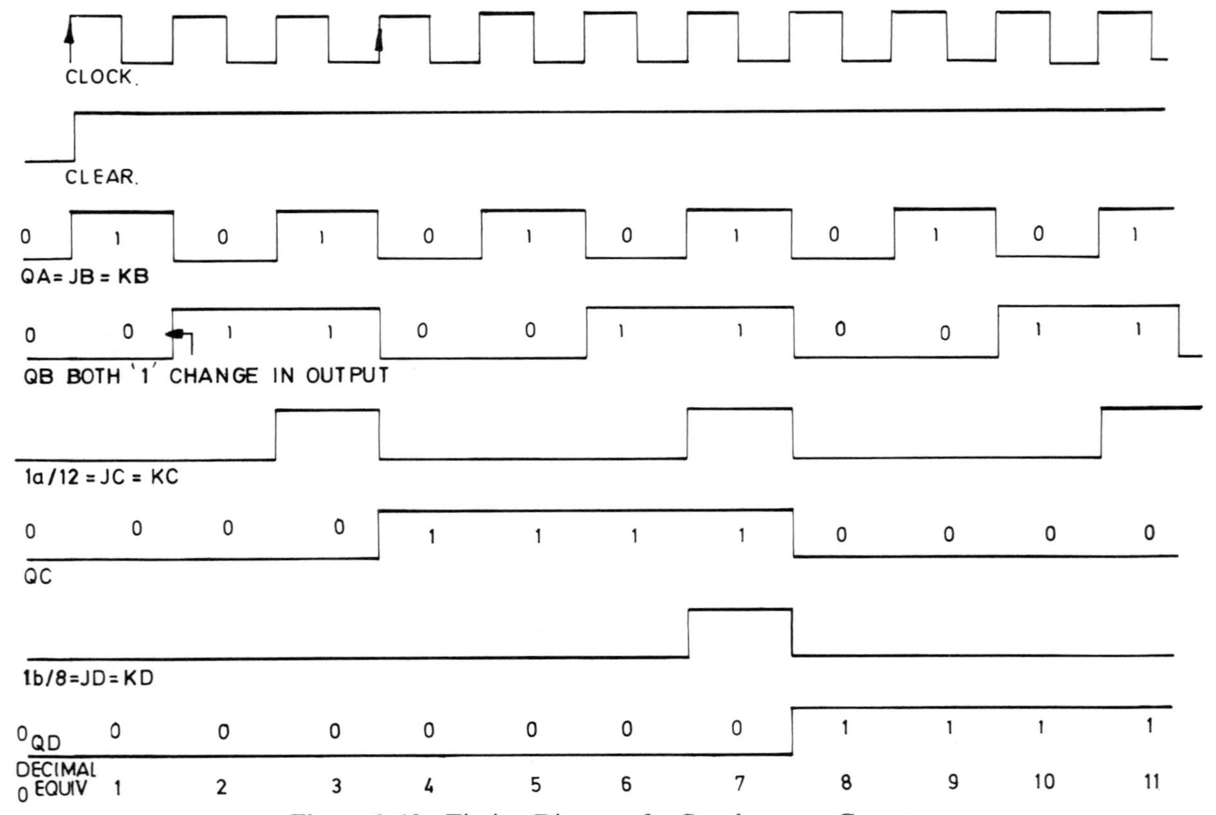

Figure 6–12. Timing Diagram for Synchronous Counter

The gating is used so that two 0s are placed on the J and K inputs of a flip-flop when no change is required in the outputs at the clock time and two 1s when it is required to be toggled in order that the correct binary sequence, in this case, is generated.

For example, J_C and K_C remain at 0 until the clock time just before bistable C needs to be toggled. When the next clock pulse is applied, J and K are both 1, causing the output to toggle. Thus J_C and K_C change from a 0 to a 1 once every four clock cycles, as required by the binary sequence; also, J_D and K_D change once every eight clocks, causing the necessary change in Q_D once every eight clock cycles.

If a nonsequential counter is desired, that is, one that does not follow the normal binary pattern, then this may be achieved by altering the gating circuitry.

Synchronous-Counter Design

Several methods are available for counter design. Some, for example, use Boolean algebra, others a more practical means that is simpler for the non-algebraically oriented person. This section describes a method that combines a knowledge of the truth table for JK flip-flops, of simple Karnaugh maps, and of the desired result without becoming involved too deeply with Boolean algebra.

What must be determined is the bit sequence that has to be produced from the counter and the signals that must be applied to the J and K inputs of all the stages, at each step, in order for the required next state to be produced. Table 6–1 is a modified truth table for the JK bistable, which lists the next-state possibilities together with the inputs required to produce them.

Table 6–1
Modified Truth Table for a JK Bistable

J	*K*	*Current State (Q)*	*Next State (Q)*
0	X	0	0
1	X	0	1
X	0	1	1
X	1	1	0

In order to demonstrate this method an example will be chosen.

Example 6–1

Design a three-stage binary sequence counter, that is, a modulo-8 counter.

Solution:

Step 1: Draw a truth table for the desired counter.

A	B	C	*Decimal Count*
0	0	0	0
0	0	1	1
0	1	0	2
0	1	1	3
1	0	0	4
1	0	1	5
1	1	0	6
1	1	1	7

Step 2: Construct a Karnaugh map that shows the direction of the changes in state (transition diagram), as follows:

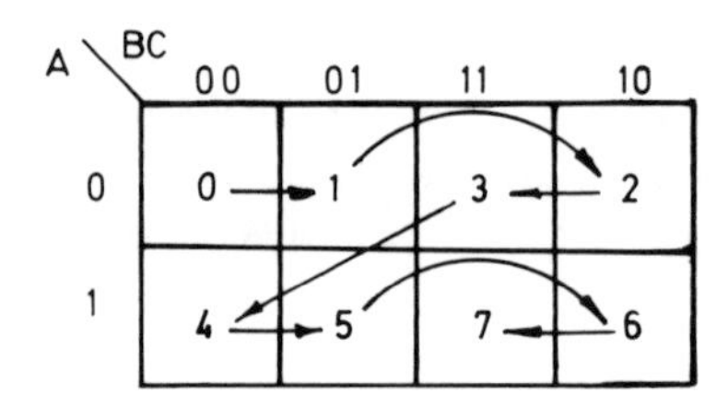

Step 3: Construct a Karnaugh map for each J and K input, such that each cell corresponds to one counter state, as in step 2. The J and K inputs required to ensure that the current state leads correctly to the next must then be entered into the correct cell location.

JC MAP

A \ BC	00	01	11	10
0	1	X	X	1
1	1	X	X	1

KC MAP

A \ BC	00	01	11	10
0	X	1	1	X
1	X	1	1	X

For the J_C map, the first state is where C = 0 and it goes to a 1. Referring back to table 6–1 this may only be achieved if J = 1 and therefore a 1 is entered in the minterm-zero (m_0) cell position. Similarly, for the next state the C output must undergo a transition from a 1 to a 0 and in this case (referring back to table 6–1) the next output state is not dependent on J_C but on K_C; hence a don't care is inserted in this position. For the transition between a decimal count of 2 and of 3, the C output must toggle from a 0 to a 1. This is a repeat of the transition between a decimal count of 0 and of 1, and therefore the same input on J_C must apply, that is, a 1.

The above sequence must be repeated until all of the cells in the K-map for J_C are covered. The process is repeated for the K_C input and then for all of the other inputs to form the following K-maps.

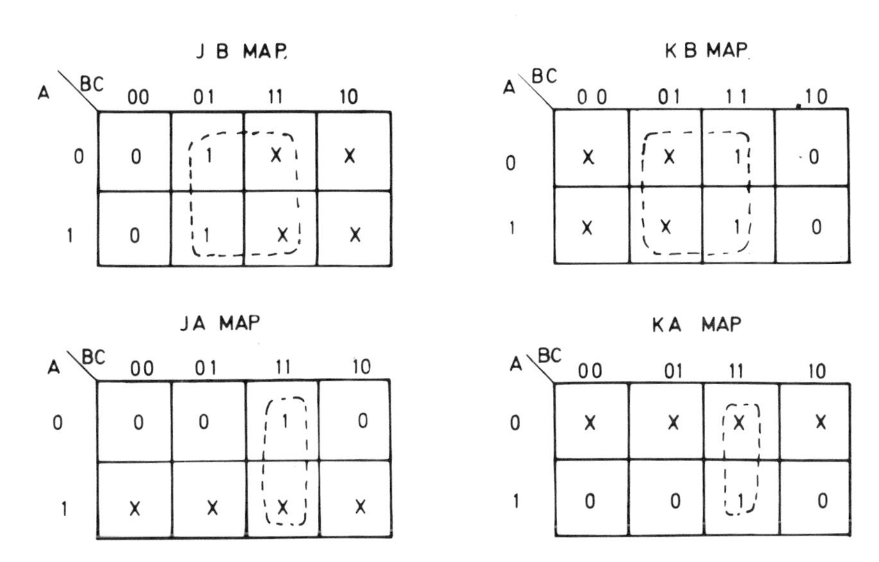

Step 4: Using the looping procedure described in chapter 3, a minimum Boolean expression for each input is obtained (remembering that don't cares may be included within a loop as a 1). From the loops already drawn on the K-maps for this example it may be seen that the corresponding Boolean expressions are:

1. $J_C = 1$.
2. $K_C = 1$ (1 and 2 are as would be expected for the C output to toggle at each clock pulse).
3. $J_B = C$.
4. $K_B = C$.
5. $J_A = BC$.
6. $K_A = BC$.

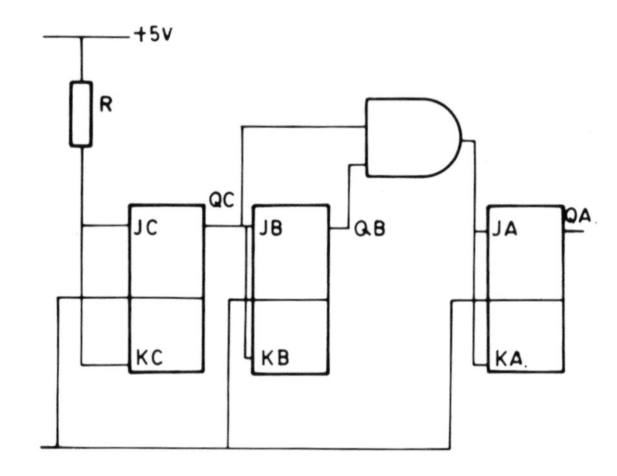

Figure 6–13. A Modulo-8 Counter

Step 5: From the Boolean expressions the counter may now be constructed as shown by figure 6–13. It may also be seen from this that it is in fact identical to the first three stages of modulo 16 counter given in figure 6–11.

Exercise 6.1

1. Design a modulo-13 asynchronous counter using four JK bistables and a NAND gate.
2. a. Design a bcd asynchronous counter.
   b. Convert this counter to give a Gray output by connecting further gating to it.
3. Design a synchronous circuit that will go through the following states: 001, 011, 110, 100, 001.

7 Combinational Logic

Binary Adders

Chapter 6 showed how a computer can subtract numbers when only an adder circuit is available by using the two's-complement number system. Obviously, with any computer a large amount of arithmetic operations will be required. These arithmetic operations may be addition, subtraction, multiplication, or division; and they may all be achieved using just an adder circuit. Subtraction, as we have seen, is the addition of the complement; multiplication is repeated addition; and, finally, division is repeated subtraction.

Therefore, the heart of any computer must be a circuit capable of addition: a *binary adder*. How these adders can be constructed and what different types of adders are available will be shown in this chapter.

The Half Adder

A very simple adder to add two 1-bit numbers together can be constructed from basic gates. Figure 7–1 shows an example of such an adder constructed from NAND gates, although the same function may be achieved using other

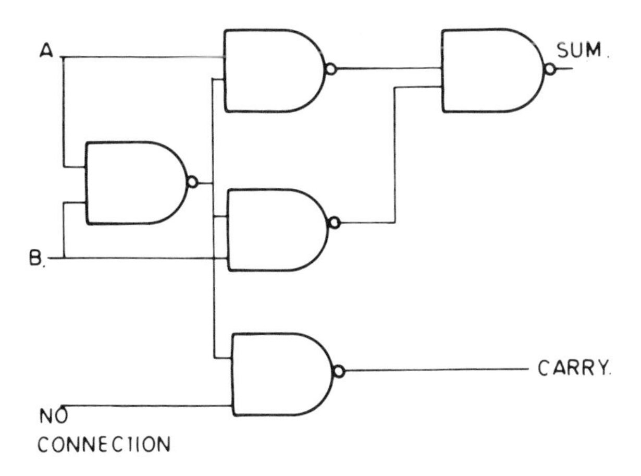

Figure 7–1. Binary Half Adder

A	B	SUM	CARRY
0	0	0	0
0	1	1	0
1	0	1	0
1	1	0	1

Figure 7–2. Truth Table for a Half Adder

gates. Figure 7–2 shows the truth table for the adder, which can be seen to conform to the rules of binary addition.

The Full Adder

Unfortunately, it is very rare for an addition to be as simple as adding two 1-bit numbers. Therefore, a larger adder is required. To achieve this the basic half adder can be extended by connecting two together, as shown in figure 7–3. The truth table for the full adder is shown in figure 7–4.

As circuits become more and more complex, they are normally drawn as squares or rectangles with their inputs and outputs labeled, rather than by showing all the gates included in them. For example, the full adder may be drawn as in figure 7–5.

The Parallel Adder

The serial adder (shown in figure 7–6) is very economical in logic but is relatively slow compared to the alternative method of addition, that of parallel addition. The parallel adder is shown in figure 7–7. With this type of adder all bits of two multibit numbers are added almost simultaneously. We say "almost" simultaneously because this is a "ripple-carry" adder, which means that the carry from each full adder is fed to the next stage, thereby affecting the sum output of that next stage.

The main problem with this is that the sum may have been generated independently of the carry and therefore might have to change when the carry from the previous stage is actually generated. This leads to a delay before the output is correct, owing to the propagation delays within each adder, similar to those for the "ripple-through" counter.

If the carry is fed into each stage at the same time as the A and B bits, then the outputs will occur simultaneously. It is possible to generate the carry, using extra logic, so that this is in fact the case. This type of adder is then called a "carry-look-ahead" adder or a "fast-carry" adder.

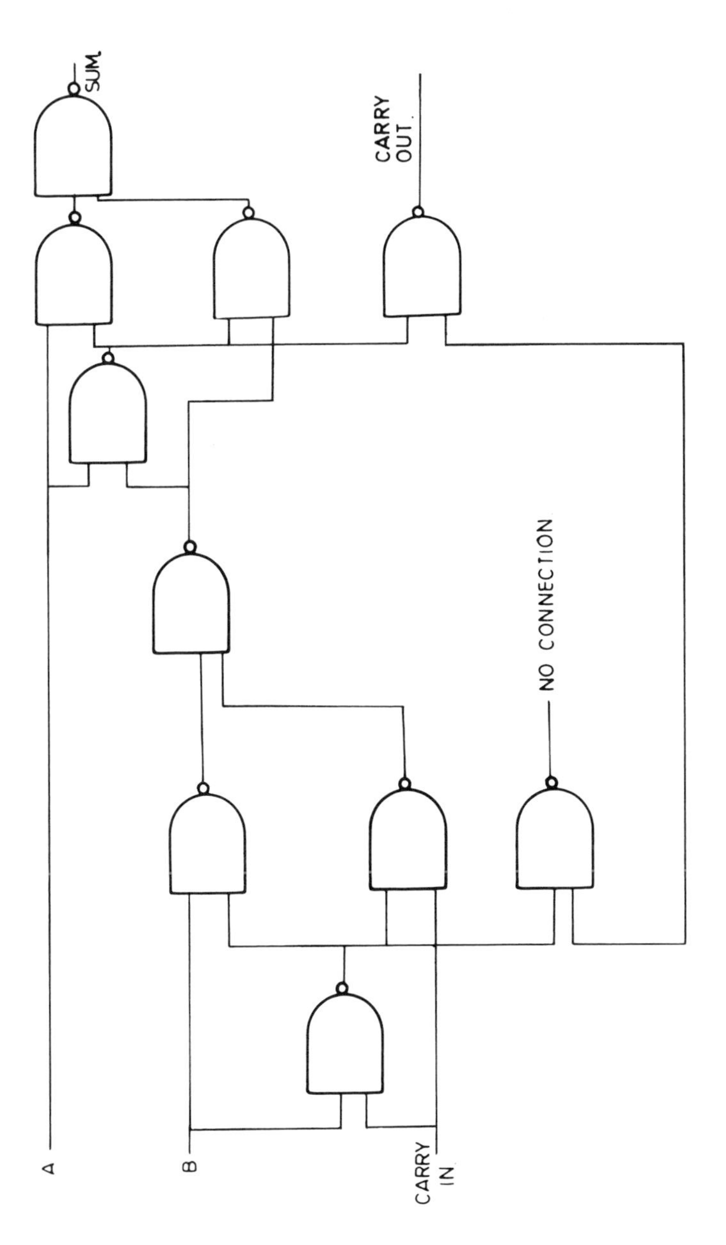

Figure 7–3. The Full Adder

INPUTS			OUTPUTS	
A	B	CARRY IN	SUM	CARRY OUT
0	0	0	0	0
0	0	1	1	0
0	1	0	1	0
0	1	1	0	1
1	0	0	1	0
1	0	1	0	1
1	1	0	0	1
1	1	1	1	1

◀ **Figure 7–4.** Truth Table for a Full Adder

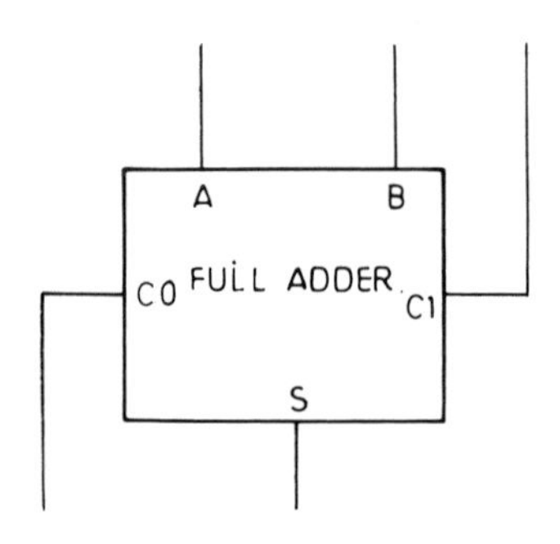

◀ **Figure 7–5.** Block Schematic of a Full Adder

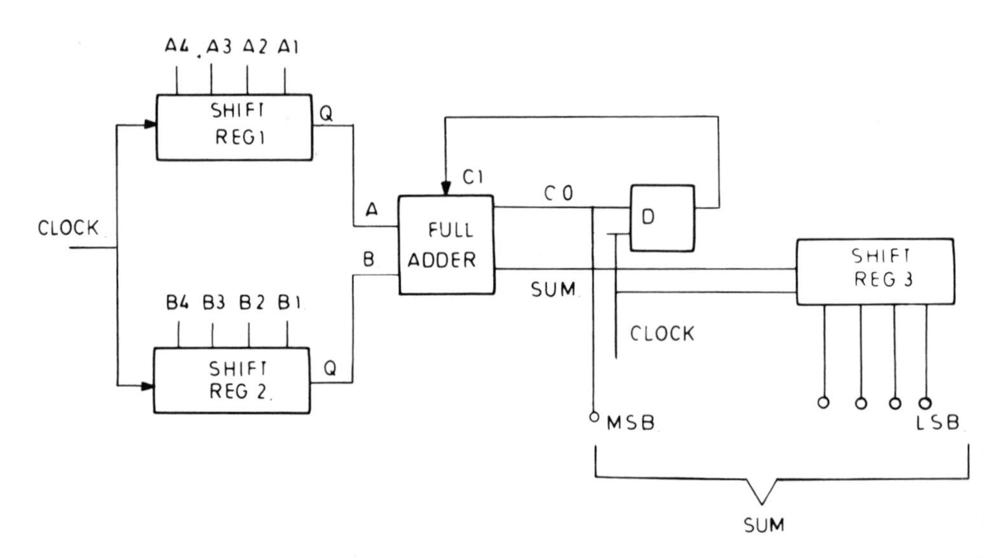

Figure 7–6. A Serial Adder Circuit

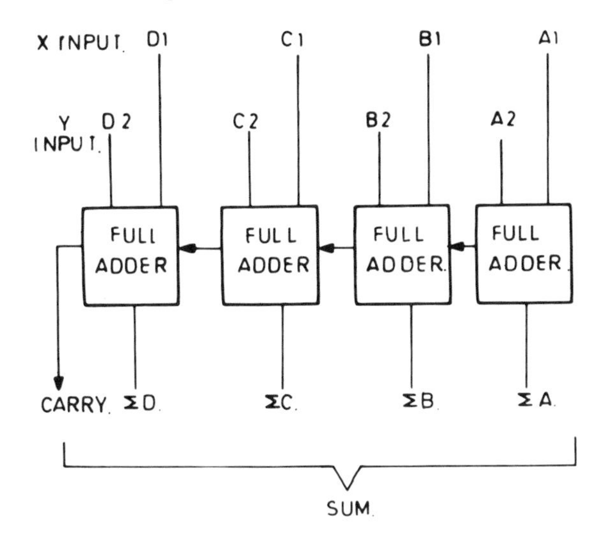

Where $X = D_1C_1B_1A_1$ and $Y = D_2C_2B_2A_2$

Figure 7-7. A Parallel Adder for the Addition of Two 4-Bit Numbers *X* and *Y*

Comparators

For many applications it is necessary not only to add numbers but also to compare them for equality or inequality. An exclusive NOR gate may be used for equality detection, since its output is a 1 when both inputs are equal. But it is only capable of comparing one bit with another. Therefore, for *n*-bit numbers, *n* EXCLUSIVE NORs with their outputs ANDed together would give the necessary equality output.

As an example, consider two 2-bit numbers A and B, where $A = A_1A_0$ and $B = B_1B_0$. In order for the numbers to be equal, A_0 must equal B_0 AND A_1 equal B_1, that is, $(A_0 = B_0) \cdot (A_1 = B_1)$. This function may be realized using the circuit of figure 7-8, which has been constructed entirely from NAND gates.

How can a circuit be designed to produce an output when two binary numbers are unequal, however? The basis for this is already in the circuit for the equality detector. Consider the logic levels of X_0 and Y_0 for the possible input conditions. If both inputs are equal then X_0 and Y_0 are both 1s. However, if $A_0 > B_0$ ($A_0 = 1$, $B_0 = 0$), then $X_0 = 0$ and $Y_0 = 1$; and if $A_0 < B_0$ ($A_0 = 0$, $B_0 = 1$), then $X_0 = 1$ and $Y_0 = 0$. Thus $X_0 = A_0 < B_0$ and $Y_0 = A_0 > B_0$.

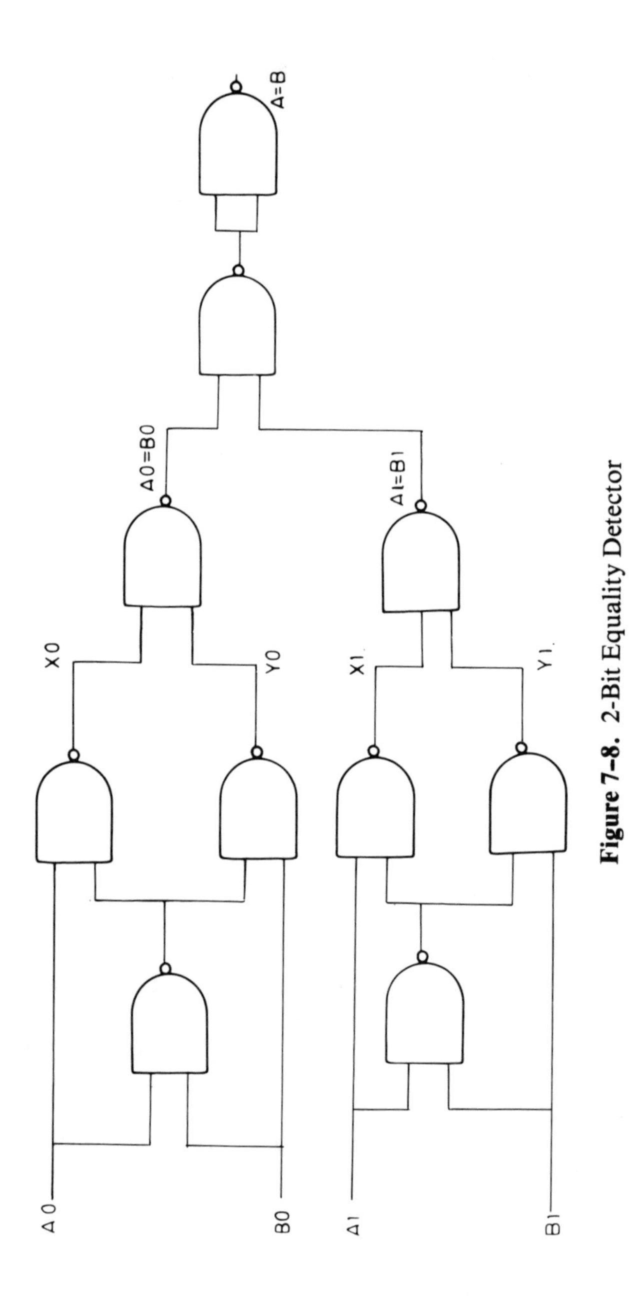

Figure 7–8. 2-Bit Equality Detector

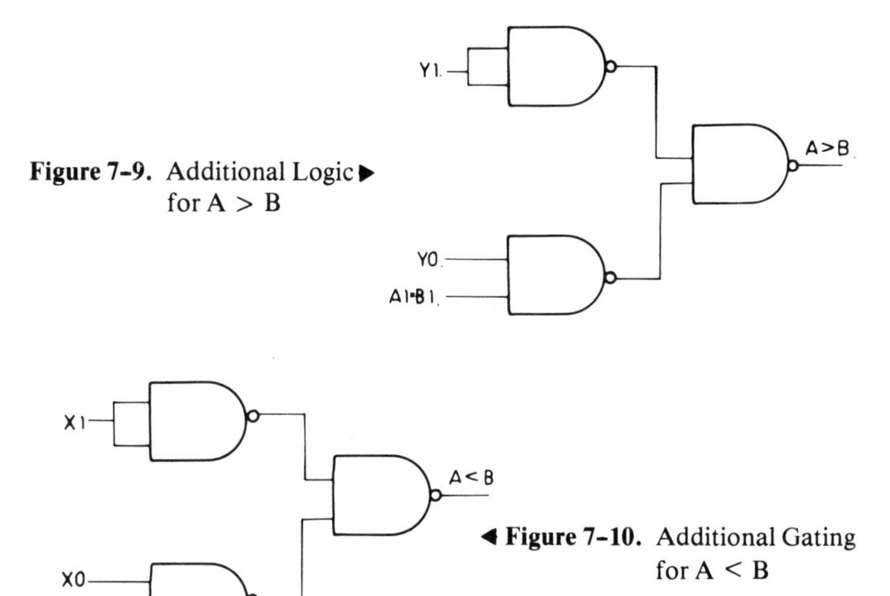

Figure 7-9. Additional Logic ▸ for $A > B$

◂ **Figure 7-10.** Additional Gating for $A < B$

Now consider once again the two 2-bit numbers. For $A > B$, then, either

$$A_1 > B_1$$

or

$$A_0 > B_0 \text{ and } A_1 = B_1$$

Therefore, by using additional gating, such as that shown in figure 7-9, the equality detector may be extended to produce an $A < B$ output.

If an output is required for $A < B$, then this condition will be true for either

$$A_1 = B_1 \text{ and } A_0 < B_0$$

or

$$A_1 < B_1$$

Hence the function $A < B$ may be produced by additional gating, such as that shown in figure 7-10.

The addition of these two circuits increases the number of gates and the circuit complexity. By utilizing the fact that $X_0 = \overline{A_0 > B_0}$ and $Y_0 = \overline{A_0 < B_0}$, a much reduced circuit may be obtained. This circuit is shown in figure 7–11 and is a modification of the equality detector of figure 7–8. It acts as a comparator as it now compares for both equality and inequality.

In practice, a 4-bit magnitude comparator is available from most manufacturers. It naturally has a complex internal structure, but this is no problem for the user. It does have one big advantage in that it is cascadable, which allows comparisons on words that are more than 4-bit. This is demonstrated by figure 7–12, which shows four 4-bit magnitude comparators being used in cascade to compare two 16-bit words. The outputs from the stage handling the least-significant bits are connected to the corresponding A > B, A < B, and A = B inputs of the stage handling the next-most-significant bits. This next stage then takes into account the result of the previous stage and its own 4-bits for comparisons.

Decoders

Decoders accept an input of a number of bits (which is normally, but not always, binary) and cause a change in state at one or more of its outputs, the outputs affected being dependent on the input code. A number of different types of decoders exist, including binary to octal, BCD to decimal, BCD to seven-segment display, and excess 3 to decimal.

Figure 7–13 is a binary-to-octal decoder, alternatively known as a 3-line-to-8-line decoder. It has three enable lines, G1, G2A, and G2B. For the decoder to be enabled, G1 must be a 1 and either G2A or G2B must be 0. Then, according to the input binary code, one of the eight outputs will be a 0, as demonstrated by the truth table (table 7–1).

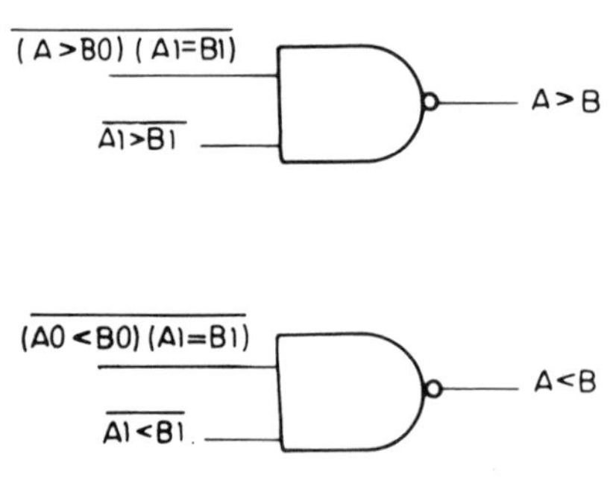

Figure 7–11. 2-Bit Comparator

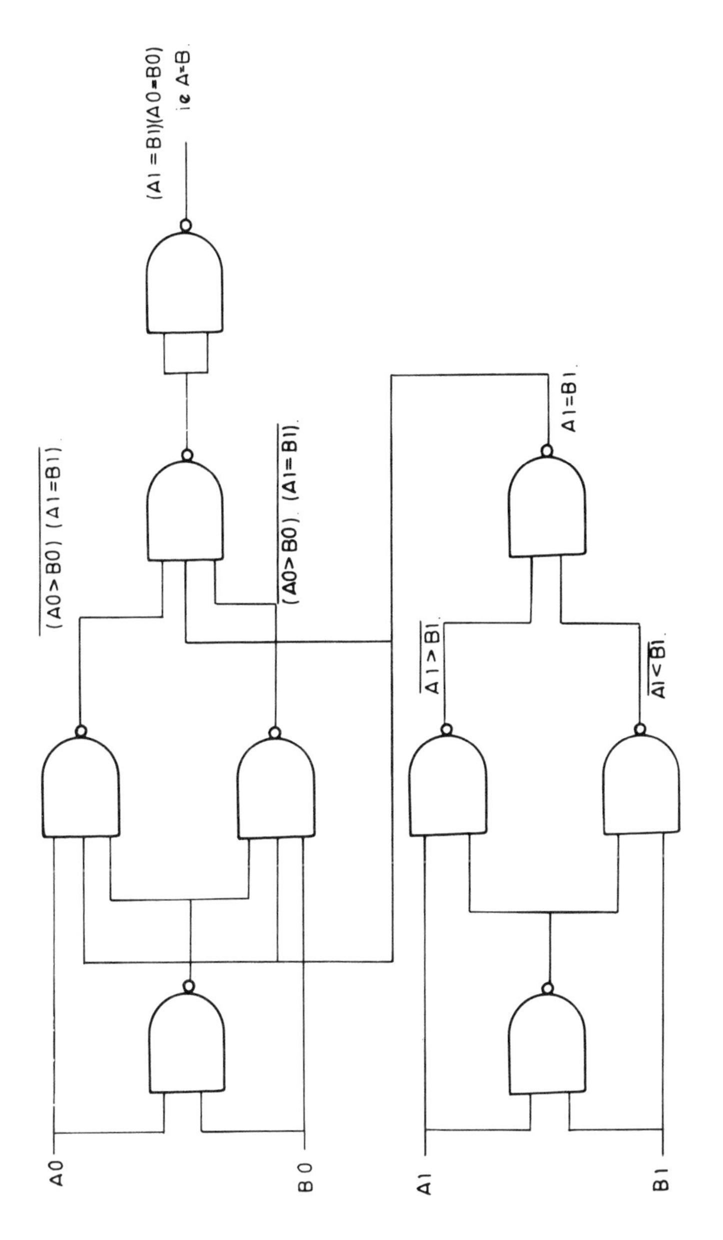

Figure 7–11 continued

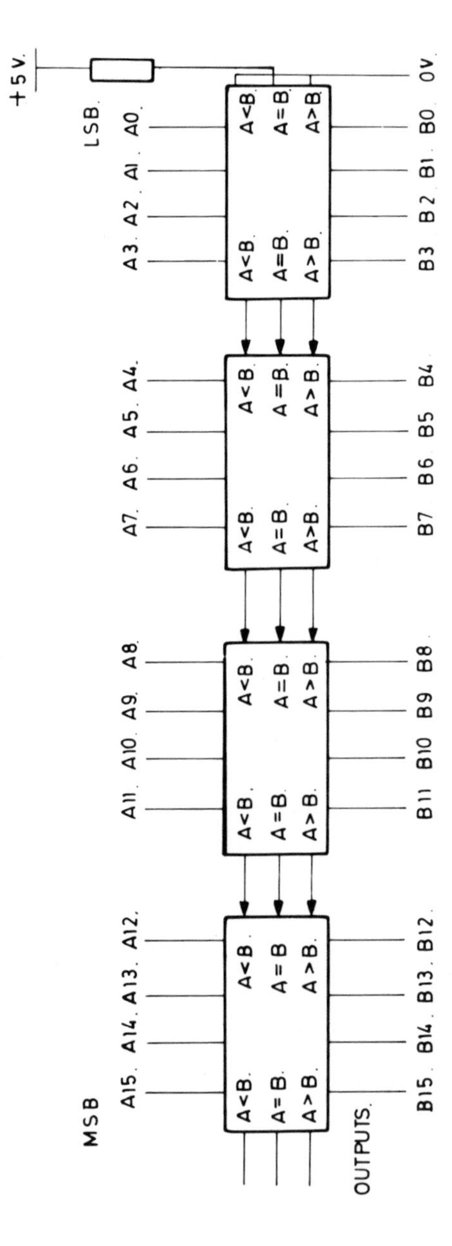

Figure 7-12. Comparison of Two 16-Bit Words Using 4-Bit Comparators

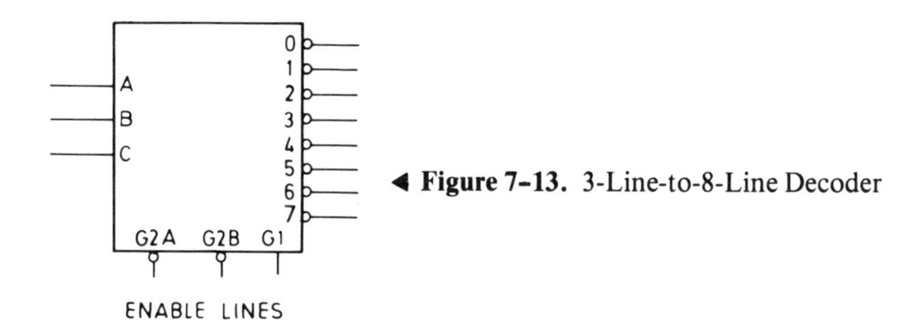

◀ **Figure 7-13.** 3-Line-to-8-Line Decoder

Table 7-1
Truth Table for a 3-Line-to-8-Line Decoder

Inputs												
Enable		*Select*			*Outputs*							
G1	*G2*[a]	*C*	*B*	*A*	*Y0*	*Y1*	*Y3*	*Y3*	*Y4*	*Y5*	*Y6*	*Y7*
X[b]	1	X	X	X	1	1	1	1	1	1	1	1
0	X	X	X	X	1	1	1	1	1	1	1	1
1	0	0	0	0	0	1	1	1	1	1	1	1
1	0	0	0	1	1	0	1	1	1	1	1	1
1	0	0	1	0	1	1	0	1	1	1	1	1
1	0	0	1	1	1	1	1	0	1	1	1	1
1	0	1	0	0	1	1	1	1	0	1	1	1
1	0	1	0	1	1	1	1	1	1	0	1	1
1	0	1	1	0	1	1	1	1	1	1	0	1
1	0	1	1	1	1	1	1	1	1	1	1	0

[a] G2 = G2A + G2B
[b] X = Irrelevant or "don't care"

Many types of equipment, such as calculators, digital voltmeters, and digital clocks, require a visual indication of the number being decoded. Light emitting diodes (LEDs) could be connected to each output to give such an indication. For example, when the binary number 011 is decoded, the LED corresponding to 3 is lit. However, this is not really an acceptable way of representing numbers, since one LED is required for each number; thus in order to represent the number 100, one hundred LEDs would be required. Therefore, the common method is to use a seven-segment display of the type shown in figure 7-14.

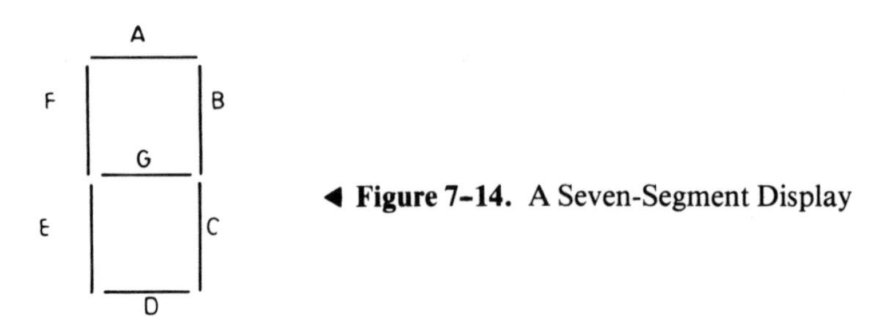

◄ **Figure 7-14.** A Seven-Segment Display

Table 7-2
Truth Table for BCD to Seven-Segment Decoder

	BCD Input				*Decoder Output to Segments*						
Decimal Equivalent	*W*	*X*	*Y*	*Z*	*a*	*b*	*c*	*d*	*e*	*f*	*g*
0	0	0	0	0	1	1	1	1	1	1	0
1	0	0	0	1	0	1	1	0	0	0	0
2	0	0	1	0	1	1	0	1	1	0	1
3	0	0	1	1	1	1	1	1	0	0	1
4	0	1	0	0	0	1	1	0	0	1	1
5	0	1	0	1	1	0	1	1	0	1	1
6	0	1	1	0	1	0	1	1	1	1	1
7	0	1	1	1	1	1	1	0	0	0	0
8	1	0	0	0	1	1	1	1	1	1	1
9	1	0	0	1	1	1	1	1	0	1	1

Since one seven-segment display can represent the numbers 0 through 9, it will be a BCD number that is to be represented. In order to light the correct segments another type of decoder must be used: a BCD-to-seven-segment decoder. The truth table for this is given by table 7-2. It may be seen from this table that for each number input, the correct segments are lit. For example, for the number 0 all segments except G are lit.

Encoders

An encoder is the opposite of a decoder. It accepts an input on one of a number of lines and converts it into a multibit code, normally binary. Typical examples are a 10-line decimal to 4-line BCD and 8-line octal to 3-line binary, shown in figure 7-15 and its corresponding truth table, table 7-3.

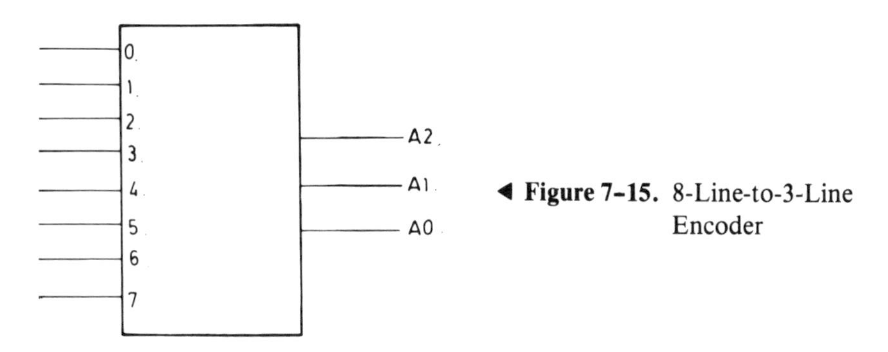

◀ **Figure 7–15.** 8-Line-to-3-Line Encoder

Table 7–3
Truth Table for 8-Line-to-3-Line Encoder

Octal In								*Binary Out*		
0	*1*	*2*	*3*	*4*	*5*	*6*	*7*	*A2*	*A1*	*A0*
1	0	0	0	0	0	0	0	0	0	0
0	1	0	0	0	0	0	0	0	0	1
0	0	1	0	0	0	0	0	0	1	0
0	0	0	1	0	0	0	0	0	1	1
0	0	0	0	1	0	0	0	1	0	0
0	0	0	0	0	1	0	0	1	0	1
0	0	0	0	0	0	1	0	1	1	0
0	0	0	0	0	0	0	1	1	1	1

Some of these devices are called *priority encoders* because they ensure that only the highest-order data line is encoded. For example, if line 8 and line 3 were both logic-level 1, then the priority encoder decodes line 8 in preference to line 3.

Multiplexers (Data Selectors) and Demultiplexers

Multiplexing is a means of selecting a specific bit from a large assortment of available bits. Only one bit may be selected at any one time, but all the bits may be selected, one after another, if required. Thus the multiplexer is the logical equivalent of the mechanical rotary switch.

Figure 7–16 shows a multiplexer that selects one of 16 input bits at a time. The specific bit selected is dependent on the binary code applied to the

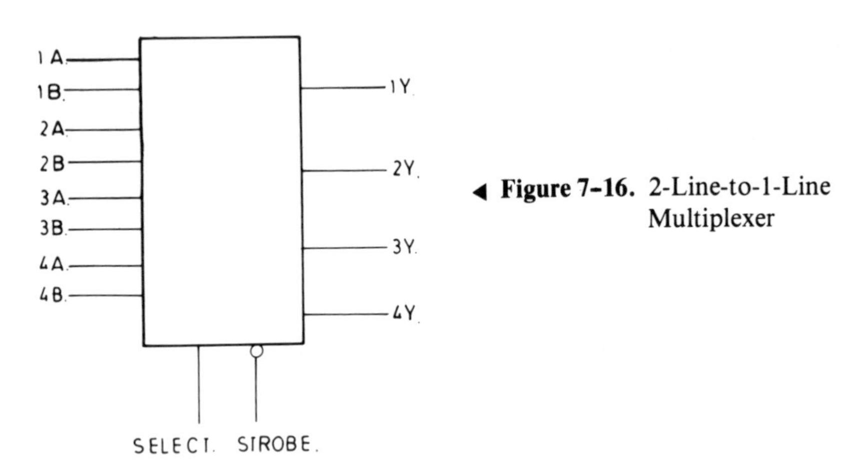

◀ **Figure 7–16.** 2-Line-to-1-Line Multiplexer

data selector inputs. For example, if input 9 is a logical 0 and the binary code 1001 is applied to the selector inputs, then the output will be a 0. If input 9 had been a logical 1, then the output would have been a 1 also.

By selecting each input in turn (done simply by connecting a binary counter to the selector inputs) the total contents of the 16 inputs can be transferred to the output. Hence the multiplexer may also be used to convert parallel data to serial.

An alternative type of multiplexer is one that accepts two *n*-bit words and selects which one is presented to its *n*-bit output. Figure 7–16 shows an example of this type of 2-to-1 multiplexer, which is used to select one of two 4-bit words. This is therefore used for routing more than one word to the same destination. In this case the multiplexer has only two control inputs, STROBE and SELECT. The STROBE line must be low to enable an output, and the SELECT line selects either the A or B inputs. If it is 0, then the A inputs are selected; if it is 1, the B inputs are selected.

Demultiplexers are the opposite of multiplexers. They select the state of a single input and present it to one of several outputs. A simple 1-line-to-4-line demultiplexer is given by figure 7–17 and its corresponding truth table (table 7–4).

This table assumes that the input is a logical 1.

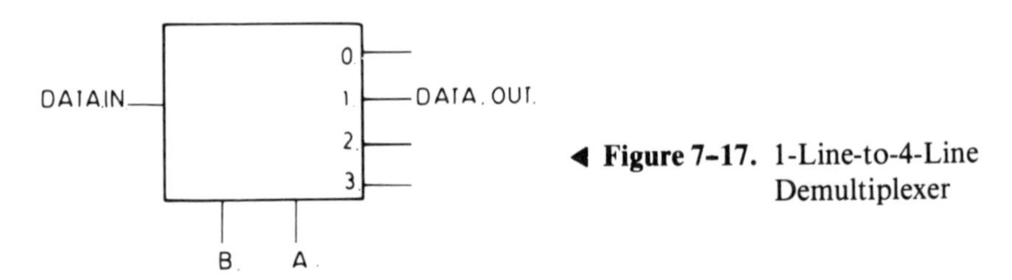

◀ **Figure 7–17.** 1-Line-to-4-Line Demultiplexer

Table 7–4
1-Line-to-4-Line Demultiplexer

	Output-Select Inputs		*Outputs*			
Input	*B*	*A*	*0*	*1*	*2*	*3*
1	0	0	1	0	0	0
1	0	1	0	1	0	0
1	1	0	0	0	1	0
1	1	1	0	0	0	1

Exercise 7–1

1. Name two types of parallel adder.
2. Name two types of device that can be used for serial-to-parallel conversion and parallel-to-serial conversion, respectively.
3. A rotary switch has its positions labeled 0 to 7. What device could be used to convert the switch position into a 3-bit binary code?

8 Memories

The power of any computer is very dependent on its ability to store and retrieve information. The larger its memory, the greater the amount of information it can process. A computer program is a list of instructions that the computer must carry out; these, along with data, are stored in memory as binary patterns of 1s and 0s.

There are a number of different types of memories on the market; the type used depends on the application, cost, and access speed required. Non-semiconductor memories include magnetic tape, cassette, disc, floppy-disc, and magnetic-core memories. Tapes and discs are used where mass storage is required for the storage of large data blocks (up to millions of words), and for main programs and external subroutines that are not immediately needed at all times but can be transferred to or from the main semiconductor memory as the need arises.

Main memory usually consists of 1K to 64K words or bytes or read-write memory, such as magnetic-core memory or, more usually now, *random-access memory* (RAM).

This relatively inexpensive main memory can be supplemented by very fast intermediate storage (*scratched memory*) in the form of extra processor registers or semiconductor memory.

The two main parameters associated with memories are *access time* and *cycle time*. Read-and-write access time is the time needed to select a memory and then read data from it or write data to it. A *memory cycle* is the minimum time required between successive accesses. The values of both of these parameters will depend on the type of memory used.

It is not the purpose of this book to enter into a discussion of external memories, such as discs and cassette, that may be connected to a computer. Only the types of memory used for a computer's main memory area and other semiconductor memories will be considered.

It must be remembered that the object of a memory is to store binary patterns of 1s and 0s, which will be interpreted by the computer's processor as either data or instructions.

Magnetic-Core Memories

Magnetic-core memories were used extensively in computers for storage before the advent of microprocessor and semiconductor memories. One of the big advantages of this type of memory is that, being magnetic, it is non-volatile, that is, it does not lose its data when its power source is switched off.

Each bit of information is stored by magnetizing a toroidal ferrite core in a particular direction, depending on whether a 0 or a 1 is required. This is done by passing a current through a wire that penetrates the hole in the core, as in figure 8–1. Thus the direction taken by the magnetic field in the core depends on the direction in which the current is passed through the wire.

The wire that passes through the core is called the *write wire,* and data may therefore be written as described in figure 8–1. However, the data also must be read, therefore, an additional wire, called a *sense wire,* must be added through the core. Information is then read by pulsing the core in the 0 magnetization direction. If a 1 has been stored in the core, the resulting flux reversal would cause an output-current pulse in the sense wire. If a 0 has been stored, no flux reversal would occur and therefore no current in the sense wire.

In practice the process is slightly more complicated, as a large number of cores must be addressed. To do this the one write wire is replaced by X and Y wires, as in figure 8–2.

To ensure that only one core is selected, half the current required to switch the core from a 0 to a 1 is sent through the X select line and the other half through the Y select line. Thus only the specific core selected (by individual X or Y lines) will have enough current through it to switch. To write a 0 to core, an additional line called the *inhibit line* is used. This is also threaded through the core. To store a 0 at B, for example, X_2 and X_1 and the inhibit line must be activated. The inhibit line neutralizes the effect of the current flowing through the select lines and thus prevents a logical 1 from being stored in core B.

The cores are now read by reversing the current in the selected lines, that is, in the 0 direction, and by monitoring the common sense line for a pulse of current. As this reading occurs by pulsing the core in the 0 direction the contents are effectively destroyed; this is one of the main disadvantages of core memories. To overcome this problem, the words read from memory are rewritten during the second half of the memory "read" cycle.

Magnetic-core memories have cycle times of between 600 and 2,000 nsec—longer than those of MOS memories. They are also more expensive than MOS memories, and therefore their popularity has decreased in recent years.

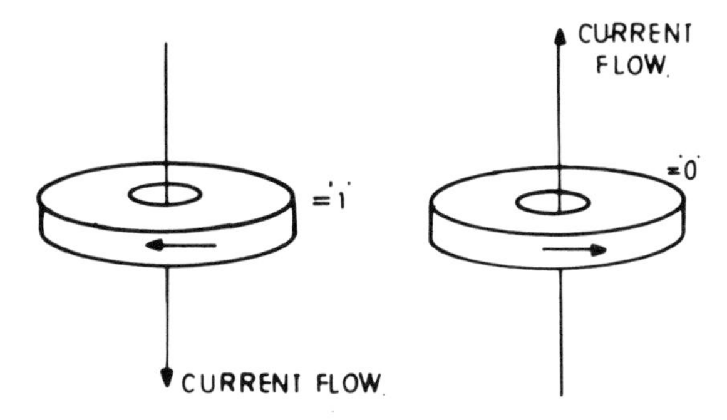

Figure 8-1. Bit Storage

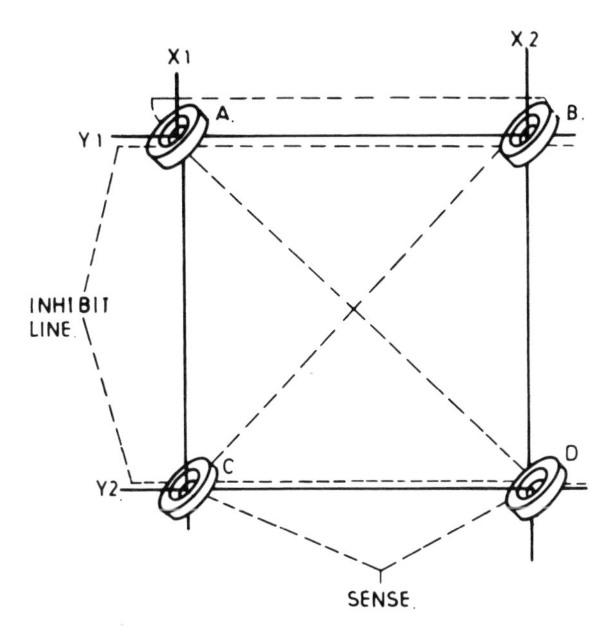

Figure 8-2. Magnetic-Core Memory

Semiconductor Memories

These are faster and cheaper than core memories and have therefore replaced them in many applications. Table 8-1 gives a comparison of the speeds of various types of semiconductor memories with that of core memory.

Table 8-1
Comparison of Memory Speed

	Access Time (nsec)	*Cycle Time (nsec)*
Core	300– 800	600–1,600
Dynamic PMOS	150– 400	250– 800
Dynamic NMOS	100– 300	200– 600
Static NMOS	60– 200	70– 300
Static CMOS	100– 150	150– 800
Bipolar	30– 100	60– 150
Masked ROM	35–1,000	70–1,500
PROM	50– 500	80– 800
EPROM	500	1,000

The properties of semiconductor memories parallel those of the corresponding logic families, and it may seem from the table that Bipolar memories—really arrays of Schottky-clamped TTL, I^2L, or ECL flip-flops with appropriate addressing circuits—are the fastest commercially available RAMs. However, their relatively high cost still restricts their use to small scratchpads and special applications that require high speed.

Semiconductor memories have the added advantage of a nondestructive readout, but the majority of them are unfortunately volatile. Probably the simplest semiconductor memory is one made up of flip-flops, such as a shift register. They are used in many applications, but because they contain many components they are more expensive and cumbersome than other semiconductor memories.

There are two main classes of semiconductor memories that supplement the register: *read-only memories* (ROMS) and *random-access memories* (RAMs). Both are two-dimensional arrays of dozens or even thousands of individual memory elements capable of storing logical 0s and 1s.

Random-Access Memories (RAMs)

Random-access memories supply the need that computers have for temporary storage of data. When they are in the "write" mode, it is possible to enter information; and when they are in the "read" mode, information can be retrieved nondestructively. Most RAMs are volatile and therefore require a battery backup in order to avoid loss of data.

There are two types of RAMs, *dynamic* and *static*. Dynamic RAMs store each data bit in the output capacitance of a MOSFET inverter. How-

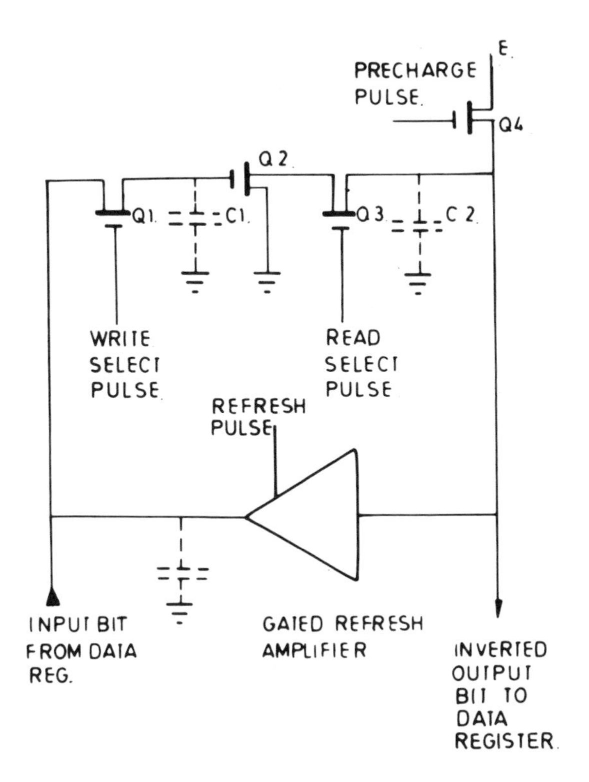

Figure 8-3. 1-Bit Cell of Dynamic MOSFET Memory

ever, this charge tends to leak away through the pn junction, and they must therefore be periodically refreshed. This is carried out by a time-shared refresh amplifier on the chip, and refresh must occur within 2 msec for most dynamic memory. This type of memory is cheaper than static and is therefore the type used for many main computer memories. An example of such a memory is given in figure 8-3.

MOSFETs Q_1, Q_2, and Q_3 in the dynamic memory form a 1-bit memory cell. The *precharge, write-select* (WS), *read-select* (RS), and *refresh* pulses are associated with consecutive phases of a four-phase clock. Each pulse, if enabled by cell-selecting logic, will transfer a bit from one parasitic capacitance to the next in a clockwise direction.

Precharge charges C_2 by turning Q_4 on and off, and WS turns Q_1 on and off. If a 1 is written, then C_1 is charged. RS turns Q_3 on and off; if a 1 is stored, the Q_2 turns on and discharges C_2 so that the inverted output is a 0. The gate-refresh amplifier inverts the selected output bit and transfers it back to its cell input approximately once every millisecond.

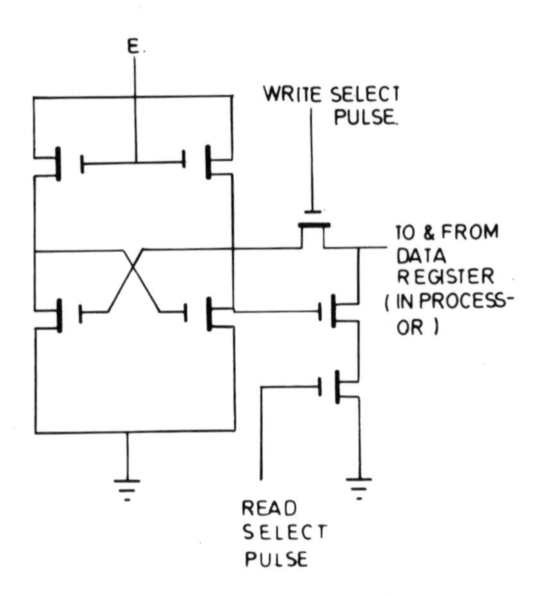

Figure 8–4. 1-Bit Cell of Static MOSFET Memory

The disadvantage of dynamic RAMs is that they require refresh circuitry, which adds to their cost. This cost must therefore be weighed against the cost of a complete memory of static RAMs.

Static RAMs have an advantage in that they do not require refreshing, since the memory cells are bistable and similar in design to conventional flip-flops. A static RAM consumes more power than its dynamic counterpart, but it requires less support circuitry and there are no problems in synchronizing the memory-refresh cycles with normal processor read and write operations. An example of a cell in a static MOSFET memory is shown in figure 8–4.

Typical Organization of Memory Chips

Generally small memories are byte-oriented, as shown in figure 8–5. Larger memories tend to be bit-oriented, as in figure 8–6. Each chip usually has address-decode circuitry and chip-select inputs so that specific memory cells can be accessed. For the byte-oriented memory 8 bits are read from each chip at a time, whereas in the bit-oriented memory only one bit at a time is read from each chip, so that 8 chips are required for a byte (8 bits) of information.

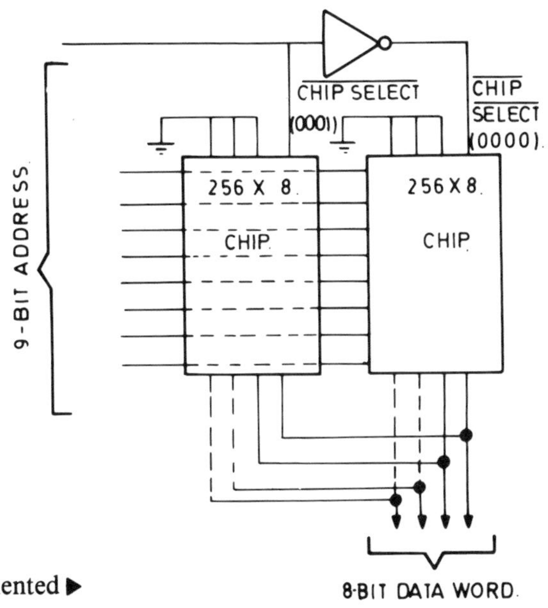

Figure 8–5. A Byte-Oriented ▶ Memory (500 × 8-Bit Memory)

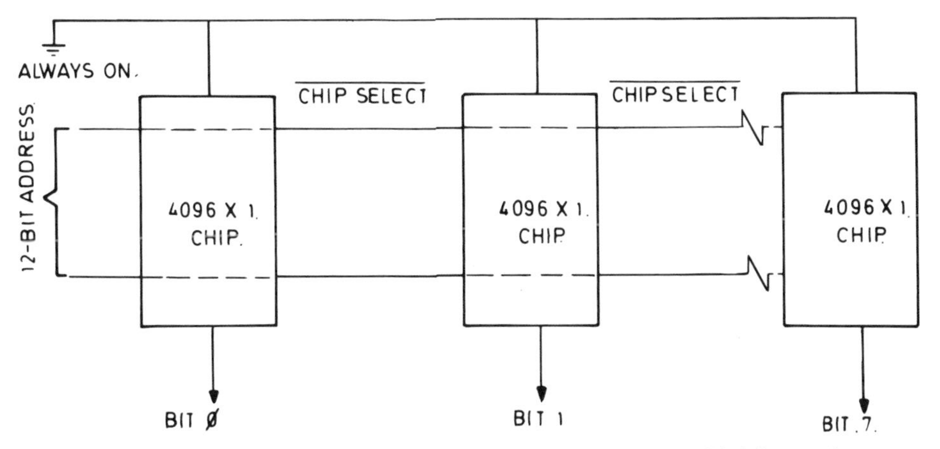

Figure 8–6. A Bit-Oriented Memory (Larger 4,096 × 1-Bit Memory)

Parity

An error in a computer's memory can cause problems in its operation. It is therefore desirable for the computer to have a means of checking for errors.

One of the simplest means is by the addition of a parity bit to every block of 0s and 1s written to memory. Extra memory chips are then included in the system to accommodate these parity bits.

The parity bit is generated according to the number of 1s in the binary word, which may be either odd or even. For odd parity the parity bit is set to a 1 or a 0 such that the number of 1s in the entire word (including the parity bit) is odd. The following are examples of odd parity:

Data	*Parity*	*Total Number of 1s*
0 1 0 1 1 0 1 1	0	5
1 0 1 0 0 0 0 0	1	3
1 0 0 1 1 0 1 0	1	5
1 0 0 0 0 0 1 1	0	3

For each of the examples the corresponding parity bit would be written to memory. On reading from memory the parity bit is reconstructed (from the data read) and compared with the parity read from memory. If they do not compare, then this indicates that there is an error in memory. Most errors in a memory are single-bit, and therefore parity is in general a good check. However, if two errors occur they will remain undetected.

Even parity may also be used in a system. In this case the parity bit is set such that the sum of 1s, in total, is even. This is demonstrated by the following examples:

Data	*Parity*	*Total Number of 1s*
0 1 0 1 1 0 1 1	1	6
1 0 1 0 0 0 0 0	0	2
1 0 0 1 1 0 1 0	0	4
1 0 0 0 0 0 1 1	1	4

Read-Only Memories (ROMs)

As with RAMs, read-only memories consist of a matrix of addressable cells. In this case, however, the cells have been permanently or semipermanently set to either 1 or 0. Thus they are nonvolatile and therefore their contents are not lost when power is removed.

There are a number of different types of ROM available on the market at present: mask-programmed ROMs, user-programmed ROMs (PROMs), erasable PROMs (EPROMs), and electrically alterable ROMs (EAROMs). All ROMs consist of crossbar matrices whose crosspoint connections may

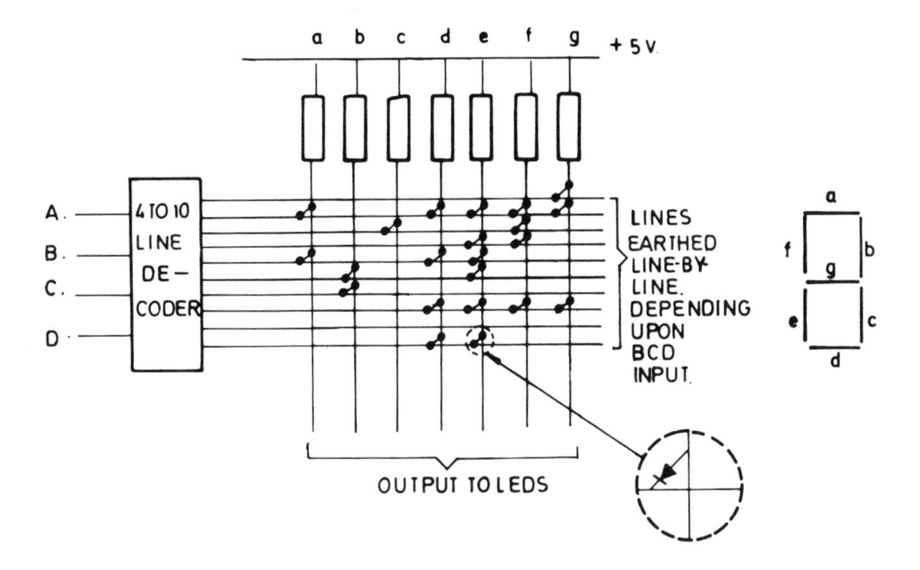

Figure 8–7. A ROM Constructed from Diodes

be conductive links, diodes, transistors, or MOSFET inverters. The storage pattern is established by selective erasure of cross-point connections during manufacture (as with mask-programmed ROMs) or before use (with PROMs, EPROMs, and EAROMs). A diagram to demonstrate how a single ROM could be constructed to drive a seven-segment display from a BCD input is given in figure 8–7. For this example the diodes are removed to store a 1 and left in place to store a 0.

A mask-programmed ROM is one that is supplied by the manufac- turers with a built-in truth table. In general these are only used where large numbers of them are required, since their cost is quite high. Typical applications for mask-programmed ROMs are in character-generation circuitry for visual-display units and calculators. Thus although they are ideal for system production, they are impractical for prototyping where their contents may require modification.

The second type of ROM, field-programmable ROM (PROM), overcomes this problem to a certain extent because it may be loaded with the specific bit pattern the user requires. Each bit position in the ROM consists basically of a fusible link. To load a logical 0 into a particular position, a surge of current is applied to the appropriate connection, "melting" or "blowing" the fusible link. This programming is carried out using a keyboard or tape-programmed fixture called, appropriately, a PROM *programmer.*

The disadvantage of both the mask-programmed ROM and the PROM is that there is no way of changing the truth table they contain. It is possible to make a limited number of changes to a PROM since logical 0s can always be added, but unfortunately it is impossible to change a 0 back to a 1.

Both the erasable PROM and the electrically alterable ROM solve this problem. The erasable PROM is programmed by applying a reverse bias of 30 v or so to the gate or drain of the pn junctions in particular cells, switching the crosspoint MOSFET permanently off and thus storing a 1 in that location. It is then erased by exposing the PROM to ultraviolet light. EAROMs are programmed by applying voltages that are well outside the memory's working range; these large voltages define the memory's operation until they are reapplied in a different way.

Charge-Coupled-Device Memories (CCDs)

Charge-coupled-device memories, like ROMs and RAMs, are made from small silicon chips. A single chip may have thousands of individual memory elements, each capable of storing a single binary bit.

These bits are stored as an electrical charge on a small metal square at each element. A charge indicates logical 1 and absence of charge logical 0.

The CCD works like a recirculating shift register as information is stored in the memory one bit at a time until all the elements are filled. A clock signal synchronizes the bit entering operation and also recirculates the contents of the memory so that the last memory bit is fed back to the first storage element. An example of a CCD is given in figure 8–8.

CCDs have the advantage that a larger number of bits can be stored in them than in an equal area of RAM chip. This is because they employ only a single component per bit of storage, wheras RAMs use a flip-flop arrangement.

Together with another type of memory, the magnetic bubble memory, CCDs represent the most recent advances in memory technology.

Magnetic Bubble Memories

Magnetic bubble memories provide a large storage capability that is nonvolatile and has a low power consumption. The bubble memory is based on the principle that in some magnetic films the plane of magnetization is either up or down in the crystal. Areas of common magnetization join up to form long strings of common magnetization, called domains, as demonstrated in figure 8–9.

No. being entered = 11101

	0	0	0	0	0	0	0	0
1ST CLK PULSE	1	0	0	0	0	0	0	0
2ND CLK PULSE	0	1	0	0	0	0	0	0
3RD CLK PULSE								
	1	0	1	0	0	0	0	0
4TH CLK PULSE	1	1	0	1	0	0	0	0
5TH CLK PULSE	1	1	1	0	1	0	0	9
number entered recirculating through CCD	0	0	0	1	1	1	0	1

Figure 8-8. CCD Memory Operation

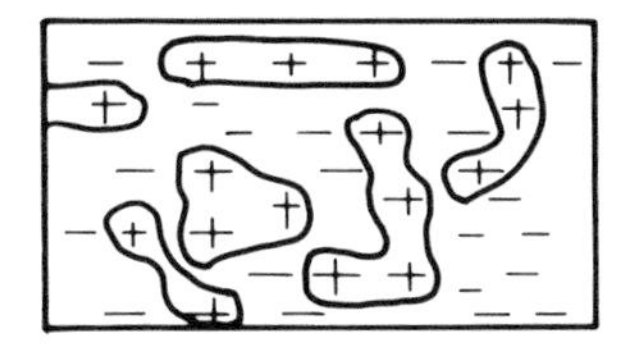

Figure 8-9. Domains

If an external magnetic field is applied perpendicular to the film, then the areas magnetized in the same direction as the magnetic field increase in volume and those magnetized in the reverse direction reduce in volume to small cylinders which appear as bubbles when viewed from above (figure 8-10).

Further bubbles can be produced by placing above the film a single "hairpin" conductor through which a current is passed. The presence of a bubble then represents a binary 1, and the absence of a bubble represents a binary 0.

A rotating magnetic field is applied in the plane of the wafer, and this causes the bubbles to shift synchronously around predefined paths in the shape of loops. The bubbles therefore act like recirculating shift registers. The paths taken by the bubbles are defined by nickel-iron elements on the surface of the film. Polarity changes in these elements are produced by the rotating magnetic field, and this causes the movement of the bubbles.

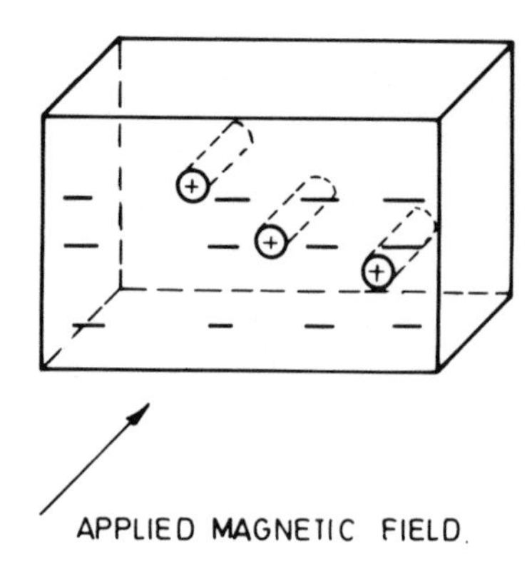

Figure 8–10. Magnetic Bubble Formations

The bubbles can be detected in one of three ways. Originally a bubble was detected using the fact that as a bubble passed under a sense wire a voltage was induced in it. However, as bubbles became much smaller the voltage they induced became too difficult to detect. Another method, which has proved more satisfactory, is to measure the change in resistance of the nickel-iron element as the bubble passes underneath it. Finally, the third method is to measure the Hall voltage produced by the bubble in an indium antimonide detector.

Magnetic bubble memories are being manufactured by the larger manufacturers such as Texas Instruments, Intel, National Semiconductors, and Rockwell. At the time of writing there are already 256-k bit chips, and even a 1-M bit chip is in existence. By contrast, the largest-capacity random-access memory currently available is 64K bits. However, it should be pointed out that in terms of speed the access time for bubble memories is longer than that for RAM, although it is faster than that of peripheral storage devices such as the floppy disc.

Exercise 8.1

1. What are the main advantages of magnetic-core memories and magnetic bubble memories?
2. State the differences between a static RAM and a dynamic RAM.
3. What are the different types of read-only memory available? Which types can be reprogrammed, and how?
4. Define the access time for memories.

9 The Digital Computer

Computers are performing an increasingly important role in our way of life. The advent of the microprocessor has meant that computer principles are now being used not only in complex electronic equipment but also in consumer products such as washing machines, games, doorbells, and children's toys. But what exactly is a computer?

A block diagram of a basic computer is shown in figure 9-1. It consists of a *storage* or *memory area,* a *central processor unit* (CPU), and an *input-output interface.*

The memory stores the instructions that will be executed by the processor. It also stores data to be used by the instructions during their executions and results of any computations carried out by the processor. Depending on the instruction it is executing at the time, the processor is capable of carrying out arithmetic functions, logical functions, data manipulation, and control functions. Thus these two units (processor and memory) make up a system that can perform calculations. However, any computer must provide a means of access so that the human operator can order the machine to execute desired tasks and can also observe the results of the tasks. Therefore, the computer also has an input-output interface with which it communicates with the outside world. This may be a *visual-display unit.*

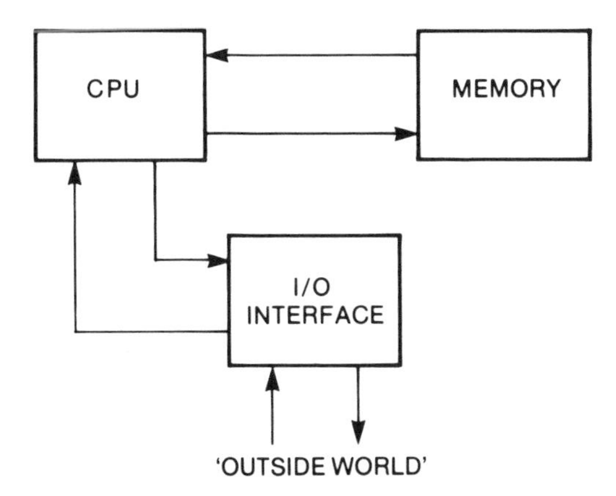

Figure 9-1. A Basic Computer

Address	Instruction
0 0 0 0 0 0 0 0	0 1 1 1 1 0 0 0
0 0 0 0 0 0 0 1	0 1 0 1 0 1 1 1
0 0 0 0 0 0 1 0	0 0 0 1 0 1 0 0
0 0 0 0 0 0 1 1	1 0 0 0 0 0 0 0

Figure 9–2. Computer Program

Instructions are stored in memory as a binary pattern of 1s and 0s, with different patterns corresponding to different instructions. Each instruction is stored in a particular location or address, and this address is also defined by a binary pattern of 1s and 0s, as demonstrated in figure 9–2.

It is obviously quite inconvenient to write the addresses and instructions as 1s and 0s, and it is therefore more common to use another number system, such as hexadecimal. The processor interrogates each address, normally in order, and fetches its contents. These are then decoded, and the function required is executed. The sequence of instructions is called the *program*, and the general term for this and its associated information is *software.* The actual electronic circuitry is called the *hardware.*

Once the processor has fetched the instruction from memory, the instruction is executed. These processes are more complex than they may at first appear. They are performed by the internal circuitry of the processor, which consists largely of devices already described elsewhere in this book. What does the processor comprise?

First, it must be able to acquire its next instruction; it must therefore know at which address the instruction is stored. Since instructions are normally stored at sequential addresses, a counter in the processor is ideal for storing the instruction addresses. Once one instruction has been fetched it can be incremented to contain the address of the next. Such a counter is called a *program counter.*

Second, the processor must be capable of decoding which type of operation is specified by the instruction; a *decoder* is therefore required. The decoder generally receives its information from a register called an *instruction register.*

Third, the processor must be capable of storing data temporarily either before or after its use in computations. Data may be stored using a shift register, sometimes referred to as an *accumulator.*

All the devices mentioned so far are those previously covered in this manuscript; counters, shift registers, and decoders. A complete processor, which includes the items discussed so far, is shown in figure 9–3.

The fourth item is the *arithmetic-logic unit* (ALU). This is really a combination of the comparator and the adder, as it can perform both addi-

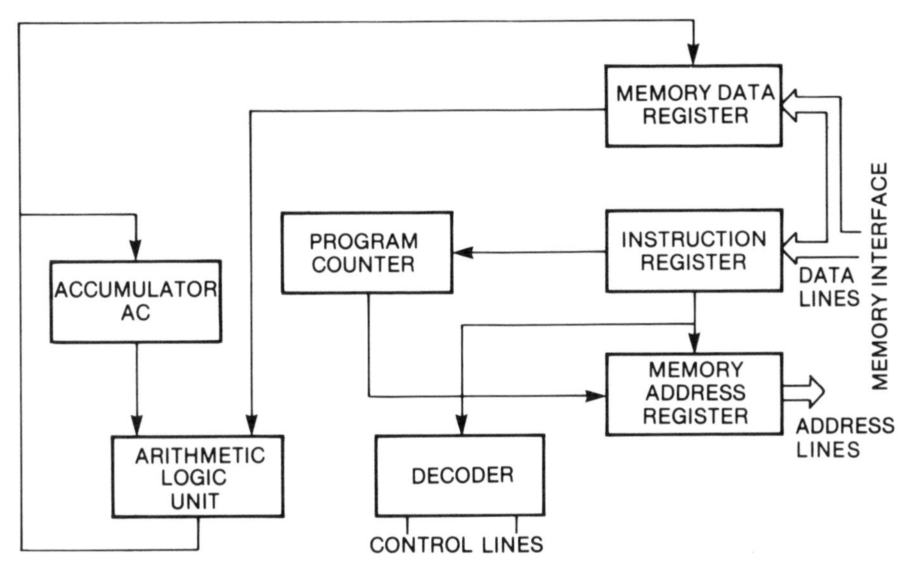

Figure 9-3. A Basic Processor

tion and comparison functions as well as a number of others. An example of an ALU is given later in this chapter.

In the processor shown in figure 9-3 there is also a *memory-address register,* which is loaded with an address from either the program counter (when an instruction is to be fetched from memory) or the instruction register. The reason that it can also be fed from the instruction register is that instructions may contain not only a code to define the function required but also an address of data in memory to be used in the instruction's execution.

As previously mentioned, a set of instructions to be executed are normally stored in successive memory locations; but it is often necessary to "branch" or "jump" to another area in memory where a different set of instructions reside. In this case the instruction contains a jump function code and also an address of the location to which the processor must jump. This address is then loaded into the program counter and causes the required jump to the new location.

The decoder (which may in practice be a ROM) supplies the control signals necessary to carry out a particular instruction. Some of the control lines are used internally and direct the actions of the registers, the program counter, and ALU. Others are used externally to direct the action of the memory of input-output devices. A typical external control line is one that indicates to the memory whether a read from it or a write to it is required.

The processor-memory interface within the computer consists of address lines, data lines, and control lines. The instructions and data in memory consist of a large number of 0s and 1s, such as 8 or 16. For fastest possible operation these are transferred to and from memory in parallel over a number of parallel wires called a *bus*. The address bits to the memory are also supplied in parallel; hence data and addresses are transferred over their respective buses, called the *data bus* and the *address bus*. If 8 bits are stored in each memory location, then a processor will normally have an 8-bit data bus and will be called an 8-bit processor.

The amount of memory (that is, the number of locations) a processor can directly address is dependent on the number of address bits it has. For an *n*-bit address bus the processor has a standard addressing capability of 2_n locations. Thus a device with a 15-bit address bus can accommodate 2^{15} or 32,768 words of memory. This is then usually written as 32-k words or 64-k bytes (where a byte is considered to be half a word). Although in practice it is possible for a processor to address more memory, this cannot be achieved without additional hardware.

Finally, the processor must be capable of transferring data to or from the input/output device via its interface. Some processors have separate data lines for this purpose whereas others utilize the data bus. A peripheral device may be addressed using the address bus as well, with one of the control lines acting as a flag to indicate that the address refers to an input-output device rather than to memory.

Most modern computers use a microprocessor as a CPU rather than one constructed from MSI and LSI. However, the basic components of the processor are the same. Also much of the necessary input-output interfacing is achieved not by discrete logic but by special microcontroller integrated circuits that can be programmed to perform specific functions or are already designed for these functions. The next section gives an example of an ALU (the basis of the processor), an example of a microprocessor, and a discussion of some of the microcontrollers available.

The Arithmetic-Logic Unit

A typical MSI arithmetic-logic unit that is available from a number of manufacturers is a 74181. It is a 4-bit ALU and as such can perform up to sixteen different arithmetic operations on two 4-bit binary words. Figure 9–4 shows the circuit symbol for the device with all its inputs and outputs labeled.

The sixteen operations are selected using the four select lines (S0, S1, S2, and S3); and these operations include addition, subtraction, incrementation, decrementation, performing shift operations, and logic operations. When the mode input (M) is high, logical operations are performed; when it

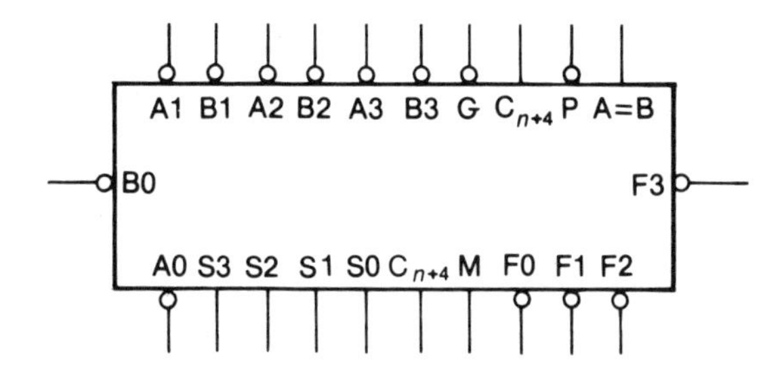

Figure 9–4. A 4-Bit Arithmetic-Logic Unit

Table 9–1
Arithmetic Logic Unit Truth Table

		Active High Data	
		M = L; Arithmetic Operations	
Selection $S_3S_2S_1S_0$	*M = H Logic Functions*	C_n = H *(No Carry)*	C_n = *L (No Carry)*
L L L L	F = $\overline{A}$	F = A	F = A PLUS 1
L L L H	F = $\overline{A + B}$	F = A + B	F = (A + B) PLUS 1
L L H L	F = $\overline{A}$B	F = A + $\overline{B}$	F = (A + $\overline{B}$) PLUS 1
L L H H	F = 0	F = MINUS 1 (2's COMPL)	F = ZERO
L H L L	F = $\overline{AB}$	F = A PLUS A$\overline{B}$	F = A PLUS A$\overline{B}$ PLUS 1
L H L H	F = $\overline{B}$	F = (A + B) PLUS A$\overline{B}$	F = (A + B) PLUS A$\overline{B}$ PLUS
L H H L	F = A ⊕ B	F = A MINUS B MINUS 1	F = A MINUS B
L H H H	F = A$\overline{B}$	F = A$\overline{B}$ MINUS 1	F = A$\overline{B}$
H L L L	F = $\overline{A}$ + B	F = PLUS AB	F = A PLUS AB PLUS 1
H L L H	F = $\overline{A \oplus B}$	F = A PLUS AB	F = A PLUS B PLUS 1
H L H L	F = B	F = (A + $\overline{B}$) PLUS AB	F = (A + $\overline{B}$) PLUS AB PLUS 1
H L H H	F = AB	F = AB MINUS 1	F = AB
H H L L	F = 1	F = A PLUS A[a]	F = A PLUS A PLUS 1
H H L H	F = A + $\overline{B}$	F = (A + B) PLUS A	F = (A + B) PLUS A PLUS 1
H H H L	F = A + B	F = (A + $\overline{B}$) PLUS A	F = (A + $\overline{B}$) PLUS A PLUS 1
H H H H	F = A	F = A MINUS 1	F = A

Source: Courtesy of Raytheon.
[a] Each bit is shifted to the next-more-significant position

is low, arithmetic operations take place, with S0 to S4 selecting the type of arithmetic or logical operation required. The various types of operation possible, for active "high" data, are shown in table 9–1.

The two binary words are applied to inputs A0, A1, A2, A3 and to inputs B0, B1, B2, B3; these then form a 4-bit output F0, F1, F2, F3 and an associated carry bit C_{n+4}.

The 74181 can also be utilized as a comparator, and it therefore provides an A = B output when two words of equal magnitude are applied to the A and B inputs. In order to perform a comparison the ALU is put into the subtract mode and its carry input (C_n) is held high. The carry output (C_{n+4}) may then be used to give information relating to inequality. For example, for "active-high" input data and for C_n being high, if C_{n+4} is low this indicates that $A > B$, and if C_{n+4} is high it indicates that $A \leq B$. In the last instance the A = B output may then be sampled to deduce whether $A < B$ or $A = B$.

It is evident from the foregoing information that the ALU is a very powerful device that can perform the function of many different types of circuits from the simplest, the AND function, to more-complex types such as addition and comparison.

Nevertheless, the ALU must have additional circuitry to provide it with the necessary control signals and data. Only then can it carry out instructions of the type used in a computer.

The Microprocessor

Until the latter half of the 1970s the majority of computer systems used a processor constructed from medium-scale integrated circuits. This bulky processor made the system itself quite large. Although such processors are still in existence, the new generation are processors on a chip and are therefore called *microprocessors.* Thus the computer's processor has been reduced from a large board to one integrated circuit less than 3 inches long and 1 inch wide.

Microprocessors are themselves a topic for a separate book, and it is not possible to provide a comprehensive description of them in this chapter. An extensive range of microprocessors is available, each type with its own particular characteristics that distinguish it from any of the others. However, the basic components are the same. In order to illustrate these, one of the most widely used microprocessors will be chosen as a typical example. This microprocessor is the Intel 8085A.

The 8085 is an 8-bit n-channel microprocessor. It is N-channel because it is constructed from NMOS materials, and it is 8-bit because it performs its operations on 8-bit data. The internal components or architecture of it are shown in figure 9–5.

It is constructed around an ALU that carries out arithmetic and logical functions on two 8-bit data blocks applied to its inputs. Control for the ALU is actually generated from the instruction decoder, which in turn

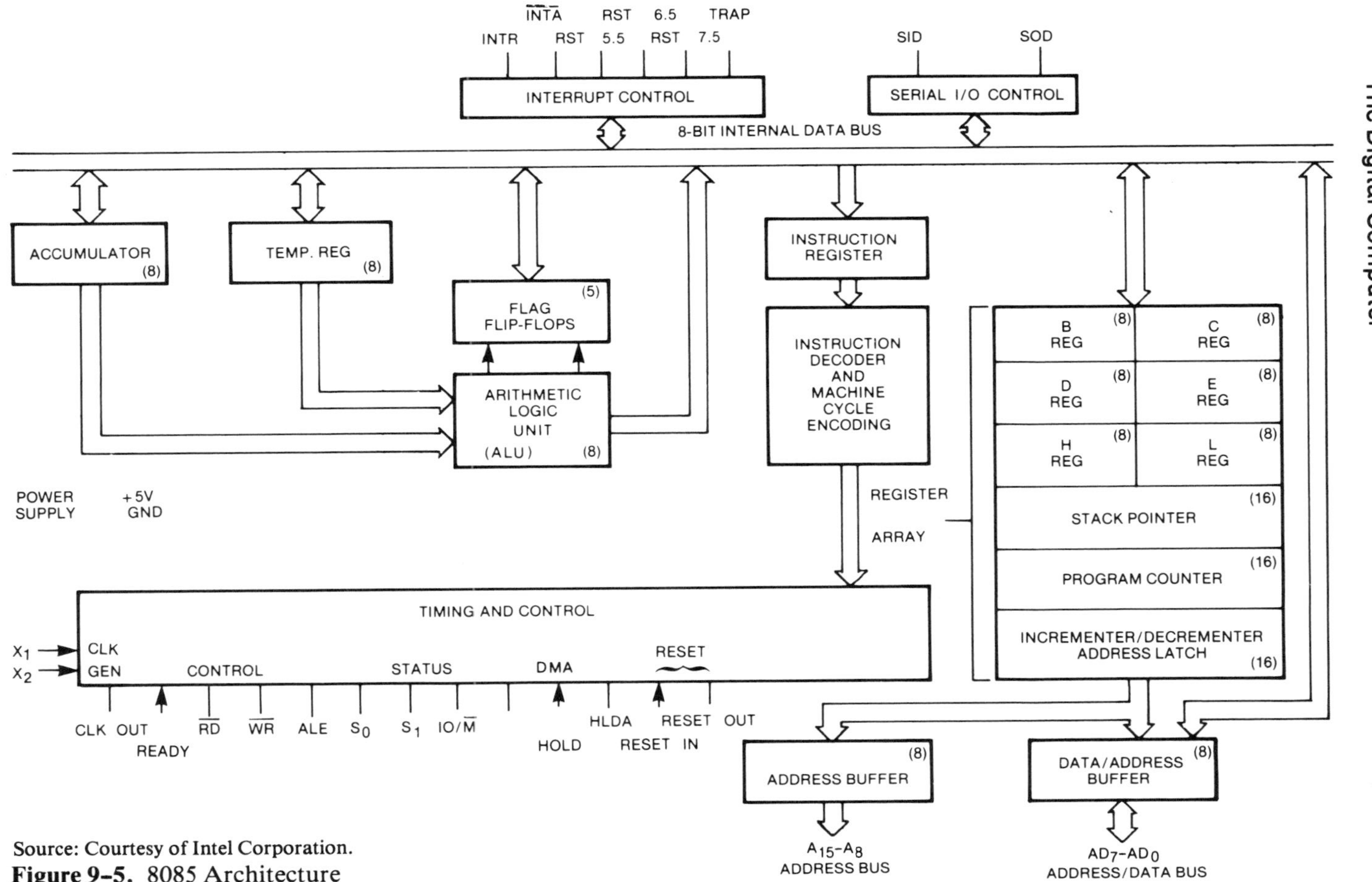

Source: Courtesy of Intel Corporation.
Figure 9–5. 8085 Architecture

receives its input from a register called the *instruction register,* which has been loaded with the instruction code fetched from memory.

Information is fetched from memory by first outputting an address to it on the microprocessor's address bus and generating the necessary read ($\overline{RD}$) control line. The memory places the information on the data bus to the microprocessor, which then receives it and loads it into its instruction register. Note that the address bus is a 16-bit bus (16 parallel lines) and that the 8 least significant bits are time-shared with the data.

The 8085 also has a register section consisting of a static RAM array organized into five 16-bit registers, the program counter (PC), the stack pointer (SP), and six 8-bit general-purpose registers arranged in pairs and referred to as B, C; D, E; and H, L. Both the program counter and the stack pointer are 16-bit registers because they refer to memory addresses. The program counter maintains the memory address of the next program instruction to be fetched from memory, and it is incremented during every instruction fetch. The stack is an area in external memory that can be used for temporarily storing data, and the stack pointer contains the address of this area. Data can be output to the location defined by the stack pointer, "pushing" data onto the stack, and the stack pointer is then decremented. Further data for the stack is then put in the next memory location down, and the stack pointer is decremented once more. This process continues until all the data required has been stored. Retrieval of this data is carried out by the microprocessor reading from the location defined by the stack pointer, "popping" data off the stack, and then incementing the stack pointer.

The six general-purpose registers can also be used for temporary data storage. Using these instead of memory may reduce the execution times of instructions.

The timing for the microprocessor is derived from a crystal applied to its X1 and X2 inputs. Instruction-execution time is therefore dependent on this crystal's frequency; a typical value for it is 6.144 MHz, giving execution times in the order of microseconds.

The remaining inputs and outputs of the timing-and-control section of the microprocessor deal with bus status, reset, input-output control, and direct memory access (DMA) control. IO/$\overline{M}$ is the input-output (I/O) control line, which selects whether the memory of an input-output device is being addressed by the address bus. Data is then transferred to or from the selected I/O device or memory.

Direct memory access is a method whereby an I/O device can address memory and transfer data to or from memory without using the microprocessor. One of DMA's uses is to transfer data between memory and a peripheral storage device such as a floppy disc. The use of DMA permits a faster transfer rate than is possible with the microprocessor executing the input-output under program control.

Data for transfer to serial I/O devices such as visual-display units. Teletypes, and cassettes may be transferred using the serial-output-data line and the serial-input-data line, but this permits a much slower transfer rate than is possible using the parallel data bus.

The final part of the microprocessor is its *interrupt control logic.* An interrupt line input to the microprocessor is used to do exactly what its name suggests: to prevent the processor temporarily from carrying out its current task and cause it to execute a new task and then to return to the original task. Hence the only detrimental effect on the execution of the original task is a time delay because of the interrupt.

As stated previously, microprocessors are being used in ever-increasing numbers and for an ever-increasing variety of applications. The main reasons for this are their low cost, small size, and greater reliability than processors made up from MSI.

Microcontrollers

A microcontroller is a device that is used to control an input-output (I/O) device in a computer, thereby relieving the central processing unit of a number of tedious tasks necessary for the input or output to occur. Originally the I/O interfaces would have been constructed from small-scale integration, but this has been increasingly replaced by microcontroller integrated circuits, which reduce system size.

The various microcontrollers available on the market include both general-purpose devices and special-purpose devices. An example of a general-purpose microcontroller is the AM 2901, which is actually a 4-bit slice processor in its own right. This device can also be used to form an 8-, 12-, 16-, or 32-bit microprocessor by cascading the devices. There are a number of support devices in the 2900 family designed for this purpose.

The alternative type of microcontrollers are specifically designed for such functions as floppy disc control, line-protocol control (used in data communication networks), CRT control, tape control, and printer control. Many of the larger manufacturers, such as Intel and Motorola, provide their own microcontroller integrated circuits, which are designed to interface with their own microprocessors. For example, Intel manufactures a 1771 Floppy Disk Controller and Motorola the MC6843 Floppy Disk Controller. These manufacturers also provide controllers for different line protocols on communication links, such as the Intel 8273 Programmable HDLC/SDLC Protocol controller, the 2652 Multiprotocol Controller and the MC6854 Advanced Data Link Controller. Other microcontrollers exist such as asynchronous communications interface adapters (ACIAs), peripheral interface adapters (PIAs), and general-purpose interface adapters (GPIAs).

It should be emphasized that in general the devices do not totally replace hard-wired logic but that they can minimize the amount of circuitry required to execute a specific function.

Exercise 9.1

1. What is meant by the term *bus*?
2. What are the functions of the:
   a. program counter?
   b. stack pointer?
   c. instruction register?
3. State the functions of which an arithmetic-logic unit (ALU) is capable.
4. What is a microcontroller and why is it used?

10 Typical Circuits

This chapter gives a number of typical applications for the logic devices considered in this book. Applications have been taken not only from the computer field but also from the secondary radar field.

Circuit 1

When one system requires data transfer to another some distance away, the transfer is generally carried out via post-office lines or radio links. It is not possible to send 1s and 0s over such links without modification, and they are therefore modulated before transmission and demodulated on reception. Modulation is a process by which the characteristics of a carrier wave are varied by the 1s and 0s. A number of types of modulation exist, but for this example differential phase-shift keying will be used. With this type of modulation a logic 0 causes a phase reversal of the carrier wave, and a logic 1 has no effect on it.

A circuit that will actually perform this function is shown in figure 10–1. it is extracted from an actual Modem circuit and is followed by filter circuitry and 600-Ω line tranformers. Figure 10–2 is the timing diagram for the circuit.

The whole circuit has three inputs: "TRANSMIT DATA" (TX DATA), "REQUEST TO SEND" (RTS), and a 38.4-kHz clock.

When the computer to which this circuit is connected wishes to transmit data, it raises its RTS line. The counters IC3 and IC4 have $\overline{\text{RTS}}$ connected to their reset inputs, clearing them and disabling any clocking action while $\overline{\text{RTS}}$ is high. As soon as RTS goes high, $\overline{\text{RTS}}$ will go low, enabling the counters, which then act as divider networks, dividing the 38.4-kHz clock to provide 2,400-Hz and 600-Hz signals. Approximately 50 msec after RTS is raised, the second counter generates a pulse, at its QD output, which clocks the D-Type flip-flop IC2A, generating READY. This delay is to allow the transmitter to become operational if using a radio link.

READY is applied to the input of the D-type IC6B, which is clocked by the 75-Hz signal derived from IC4. This causes the Q output to go high, generating READY FOR SENDING (RFS). The circuit receiving RFS then replies with the data for transmission.

Figure 10–2 shows a typical data stream. This data is synchronized to the 600-Hz clock by presenting it to the input of the D-type flip-flop IC2B.

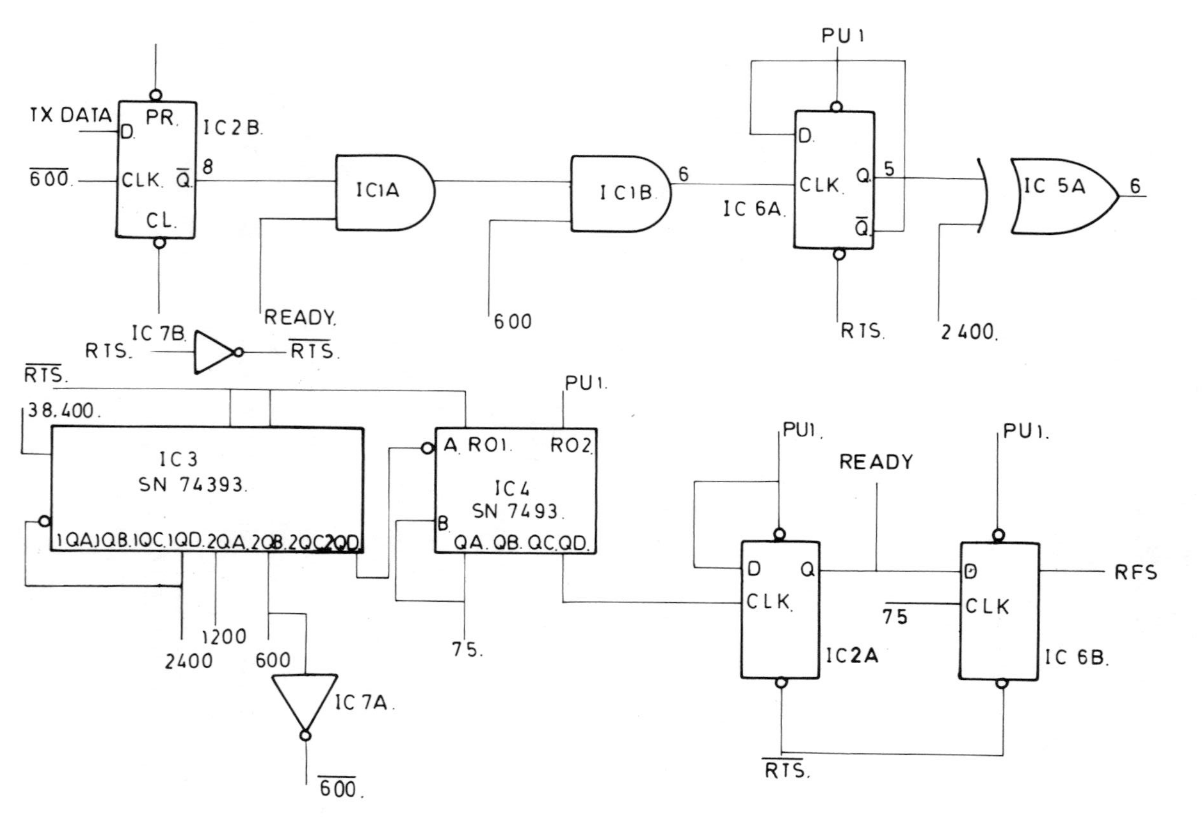

Figure 10-1. Modulator Circuit

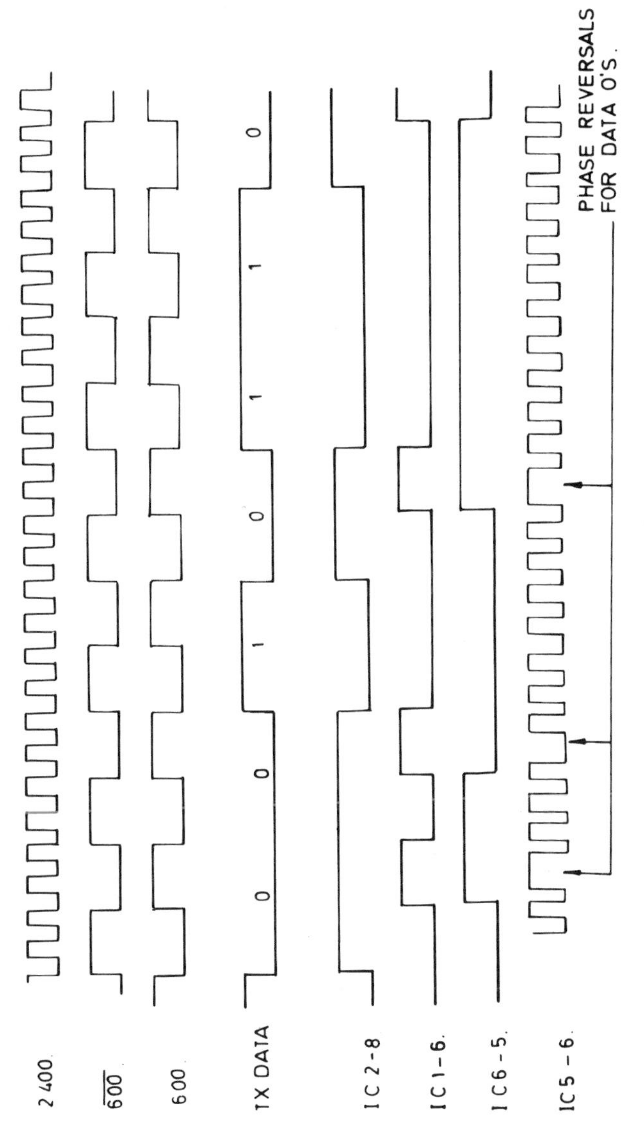

Figure 10-2. Timing Diagram for Modulator Circuit

The $\overline{Q}$ output of IC2B is then presented to IC1A, which is enabled by the READY signal. At IC1B the waveform is then used to gate the 600-Hz clock signal, and the resultant waveform is presented to IC6A, where it is divided by 2. The output of IC6A is fed to the EXCLUSIVE OR gate IC5A, together with the 2,400-Hz carrier, to produce the differential encoded data, as shown by the timing diagram in figure 10–2.

Circuit 2

Many types of electronic equipment require clock signals to carry out their functions. In order to obtain the required frequency, the clocks may be generated by a crystal oscillator that has its output divided down to the desired frequency. The following is an example of a circuit (see figure 10–3) that generates one of two frequencies, either 689.6 kHz or 1 MHz, depending on the state of an input signal called $\overline{\text{TIMER ENABLE}}$.

If $\overline{\text{TIMER ENABLE}}$ is high, then the D-type flip-flops of IC4 are enabled for clocking since their present and clear inputs are both high. However TIMER ENABLE is inverted by IC3 and this clears IC10b, IC10a, IC9a, and IC8b and presets IC9b, preventing the clock input from affecting their outputs.

Thus for $\overline{\text{TIMER ENABLE}}$ high, IC4b and IC4a both act as divide-by-two circuits and divide the 4-MHz signal, by four generating a 1-MHz signal. The 4-MHz signal is generated by the crystal oscillator constructed from TR1 and its associated components.

As IC8b has a low on its clear input, its Q output is low and its $\overline{Q}$ output is high. This high enables the NAND gate IC6c and gates through the 1 MHz resulting in an inverted 1-MHz output.

For $\overline{\text{TIMER ENABLE}}$ low, then IC4 flip-flops are disabled and IC10, IC9, and IC8 are enabled. These form a divider circuit, which divides down the 16.552-MHz output of the crystal oscillator. IC10b, 10a, and 9a are each divide-by-two circuits, giving a total division of 8. IC9b and IC8b then form a divide-by-three circuit, and therefore the overall circuit performs a divide-by-twenty-four function. Hence the output from IC8b is equal to 16.552 MHz divided by 24, or 689.6 kHz.

Because IC4a's output is a 1, the 689.6-kHz signal is gated through and inverted by IC6c to provide the "clock" output.

Circuit 3

This circuit provides a means of detecting the pulse spacing of two input pulses A_1 and A_2. If their spacing is nominally 3μsec, then a "pulse detect"

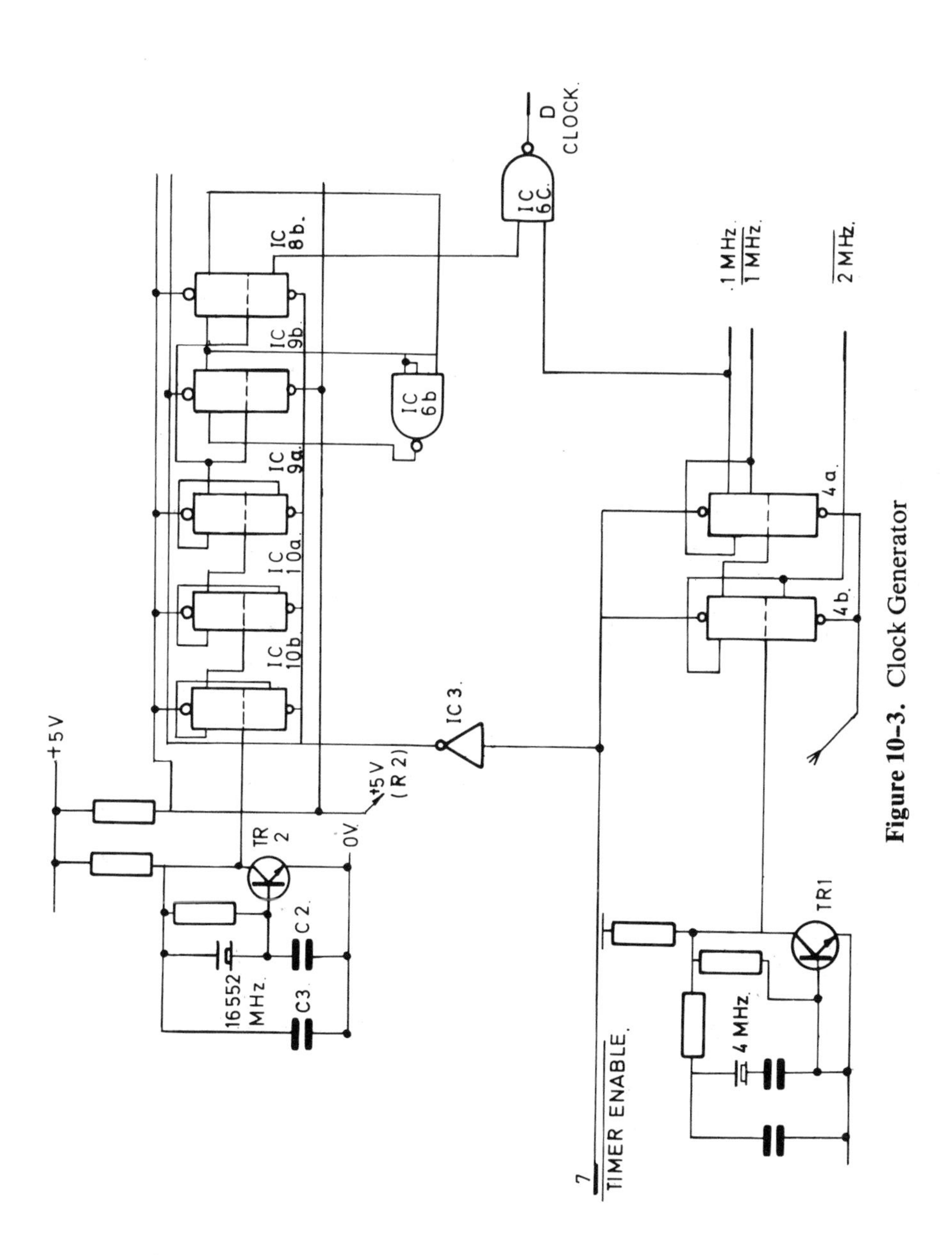

Figure 10–3. Clock Generator

output is generated. However, if the pulses' spacing is anything other than 3μsec, then no output is produced.

Figure 10–4 shows the circuit. It consists of a shift register constructed from two JK flip-flop packages, IC4 and IC5, with inverters on the inputs to ensure that J and K inputs are the inverse of each other. The first input pulse A1 is received and clocked down the register by $\overline{\text{CLOCK 1}}$, which has a frequency of 1 MHz to provide the whole number of microseconds delay required (owing to the pulse spacing being a whole number of microseconds). The delayed signal is sampled by the gating circuitry of IC1, IC2, and IC3, such that a pulse is generated at the output of IC1. This pulse is then synchronized to $\overline{\text{CLOCK 2}}$, a 2-MHz signal, using IC6a, and its output is presented to IC6b, which is clocked by the A2 pulse. The timing diagram of figure 10–5 shows the circuit's operation.

Circuit 4

Many computer systems have a circuit called a *watchdog* that causes an action in the event of their failure to execute a program. They may cause a reload of the system program, cause a changeover to a standby computer or give a visual or audible indication to the operator. The circuit shown in figure 10–6 is an example, which checks that two computers are operating and in the event of failure causes an associated light-emitting diode (LED) to be extinguished and also sounds an audible alarm.

The system programs of the two computers are designed so that one specific part of the program generates a pulse that is applied to a retriggerable monostable. It is also designed so that this pulsing program is executed at an interval of less than 1 second, keeping the monostable triggered. Therefore, if the computer does not return to the pulsing program, then the monostable will no longer remain triggered.

The two retriggerable monostables are IC1A and IC1B. While they are continually retriggered, their $\overline{Q}$ outputs will remain low. However, as soon as a retrigger pulse is not present, $\overline{Q}$ will go high, generating a positive-going edge to clock the following D-type flip-flop IC2A or IC2B. Zero volts is connected to the D-input of the flip-flop, which causes the corresponding Q output to go low when the flip-flop is triggered. Hence either QA or QB goes low, depending on which computer failed, the main or the standby. Also, the action of the monostable's $\overline{Q}$ output going high extinguishes the associated LED, indicating failure.

QA and QB form the inputs to an open-collector NAND gate IC3B. If QA or QB are low, then IC3B pin 6 is high. This is then presented to the base of transistor TR2, turning it on. Consequently its collector and hence the base of TR1 goes low, turning TR1 on, sounding the audible alarm.

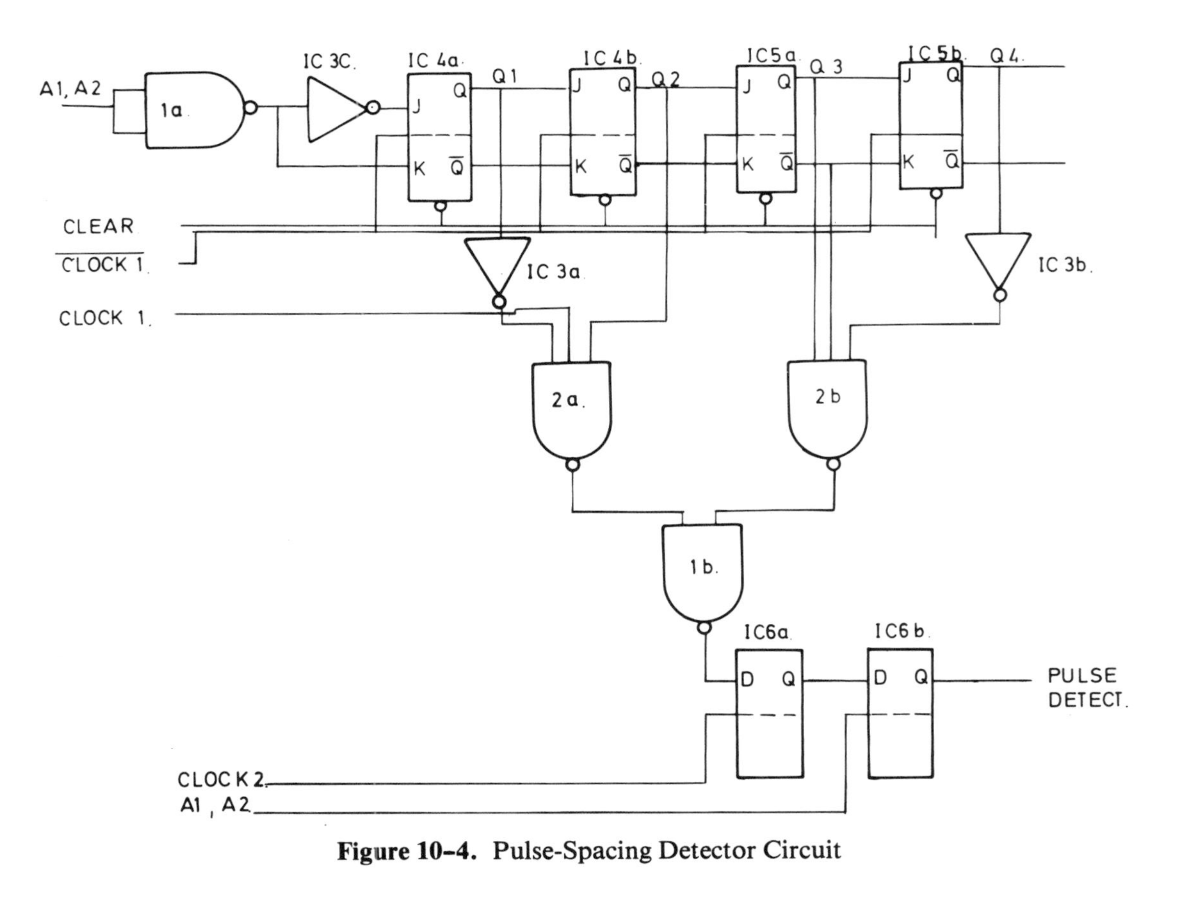

Figure 10–4. Pulse-Spacing Detector Circuit

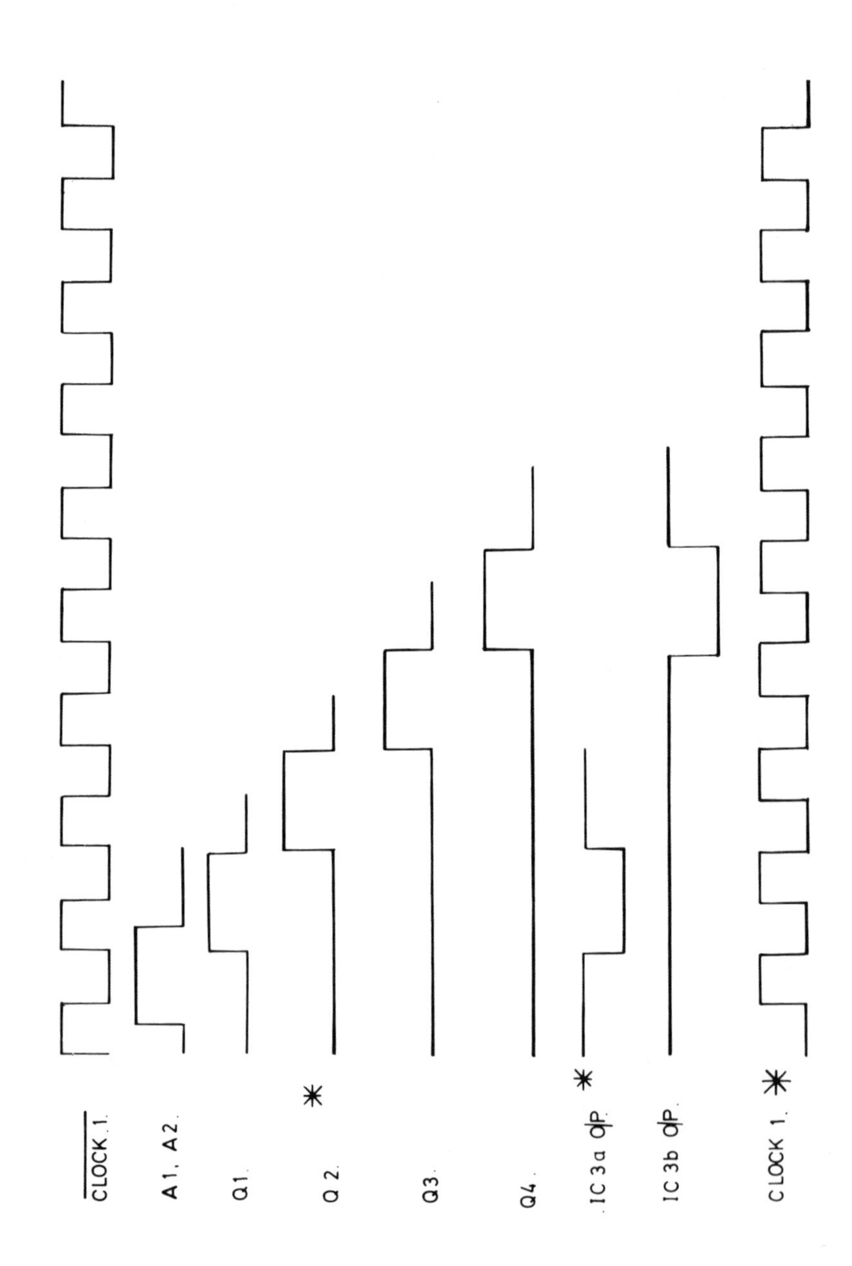
CLOCK 1
A1, A2
Q1
Q2
Q3
Q4
IC 3a O/P
IC 3b O/P
CLOCK 1

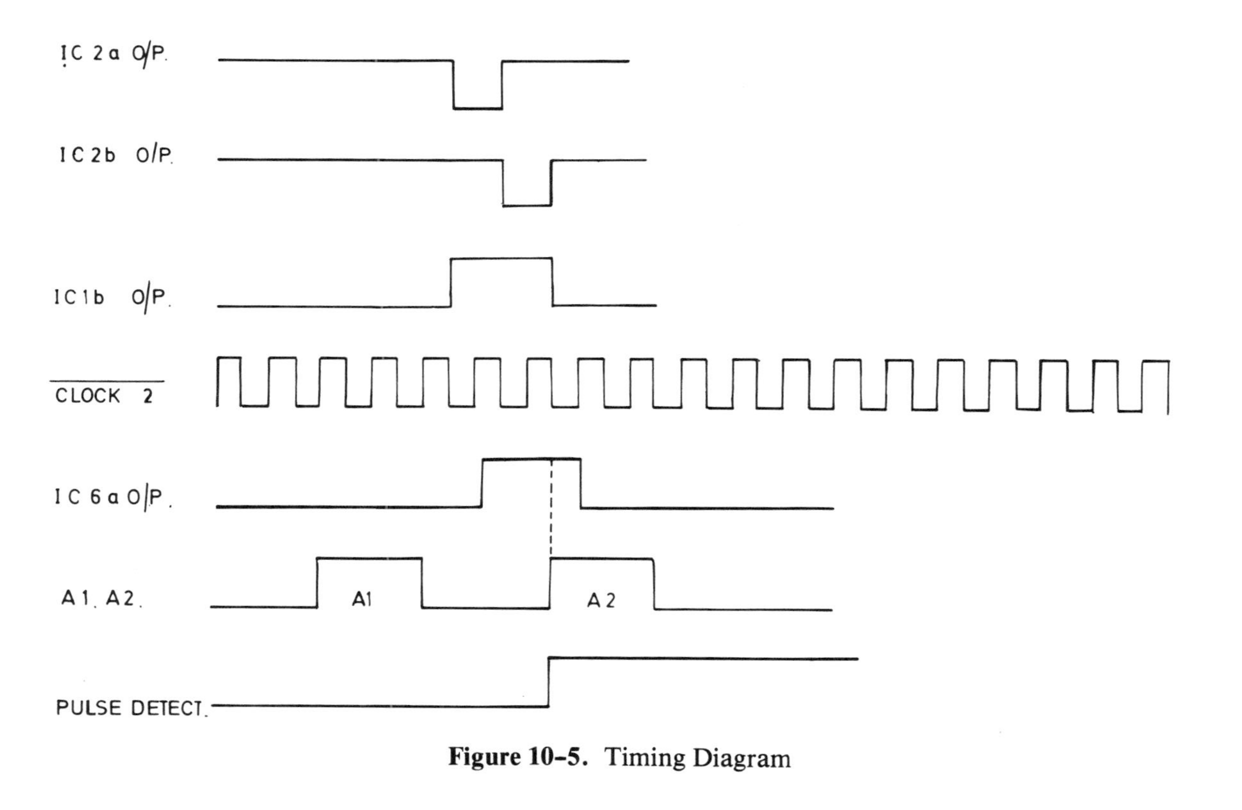

Figure 10-5. Timing Diagram

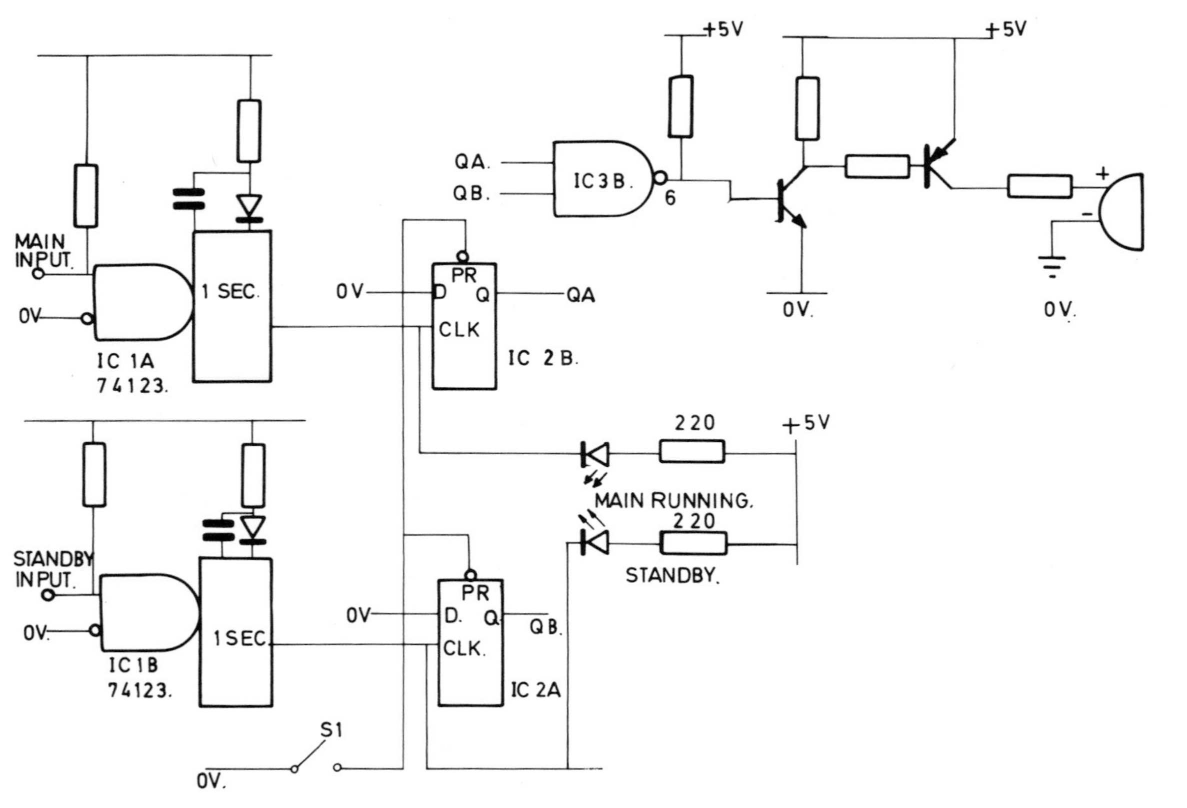

Figure 10–6. Watchdog Circuit

The alarm may then be silenced by closing switch S1. This applies a low to the presets of both D-type flip-flops, presetting both QA and QB to a 1 and clearing the alarm. Thus the alarm is cleared and the LED associated with the failed computer will not be lit until the retrigger pulses are reapplied.

Circuit 5

Figure 10–7 is part of a peripheral-interface circuit in a microprocessor-based system. It is by no means complete, but it is given as an example of the use of decoders, demultiplexers, and multiplexers.

The microprocessor used in this system is a Texas Instruments 9900 and, unlike the 8085 described in chapter 9, it has a 16-bit data bus (DO–D15) and a 15-bit address bus (A0–A14). DO and A0 are the most-significant bits of the respective buses.

The circuit's function is to allow communication between the microprocessor and the peripheral via the microprocessor's serial output (CRUOUT) and serial input lines (CRUIN) and via the 8-bit transfer bus ($\overline{TBO}$ to $\overline{TB7}$) to the peripheral. Direct memory access is also allowed by providing translation of two 8-bit data bytes into a 16-bit word, which can be output directly onto the computer system's data bus.

First, as with any input-output (I/O) device within a microprocessor-based system, the circuit must have a specific address. This is provided by IC2 and IC1. IC1 is a decoder that is enabled by the application of a low to its strobe inputs IG and 2G. This low is generated only when all the inputs to IC2 are high, that is, when A3, A5, A6, A7, A8, and A9 are all at a logic 1 and A4 is a logic 0. For the purposes of this example it may be assumed that A0, A1, and A2 are all low, and therefore the address lines A0 to A9 must form a code of 0001011111 to enable the decoder. The remaining address bits enable specific functions in the circuit.

$\overline{CRUCLK}$ is generated whenever data are present on the $\overline{CRUOUT}$ line. The line above the signal indicates that it is low when active. In order to convert the serial data on the CRUOUT line into parallel data ($\overline{TB0}$ to $\overline{TB7}$), IC4 must first be enabled by presenting an address to the circuit with A10 and A11 both low. With $\overline{CRUCLK}$ low, $\overline{EN4}$ is generated; and by incrementing the address bits A12, A13, and A14 for each serial data bit, the serial data is converted into parallel for presentation to the transfer bus..

Data input from the transfer bus, which is required to be read on the microprocessor's serial input line CRUIN, is presented to IC3, which is enabled by $\overline{EN0}$. $\overline{EN0}$ is also generated when A10 and A11 are both low; and when A12, A13, and A14 are successively incremented, each bit of the parallel data is output from IC3 on the CRUIN in turn.

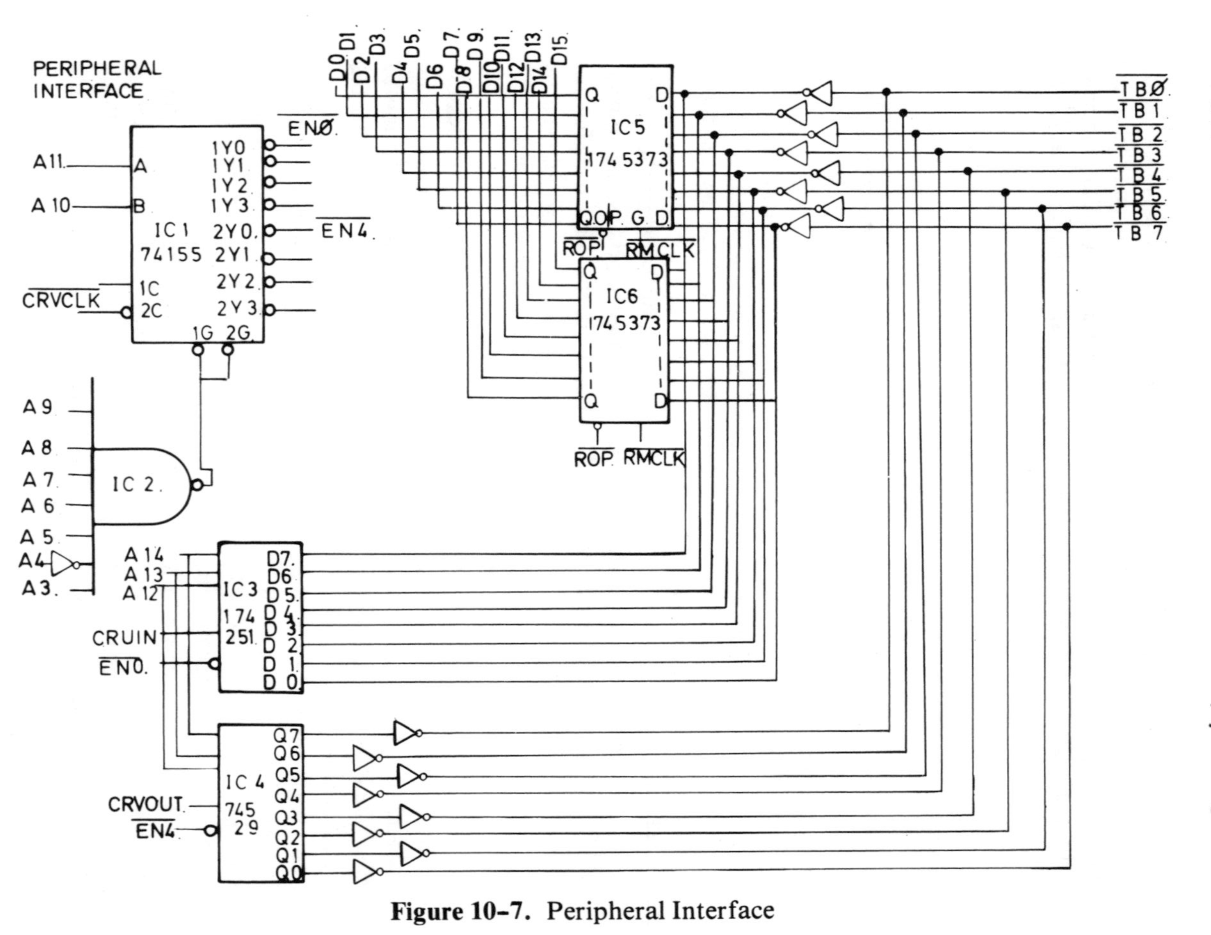

Figure 10–7. Peripheral Interface

Thus for both serial-to-parallel and parallel-to-serial conversion, the board must be provided with eight successive addresses, beginning with 0000101111100000.

IC5 and IC6 are D-type flip-flops that have 8-bit data on their inputs. $\overline{\text{RHCLK}}$ and $\overline{\text{RLCLK}}$ are then generated in turn (by control circuits not shown) such that IC5 D-types are clocked, followed by the D-types in IC6. By generating ROP, the two blocks of 8 bits are output from the D-types at the same time for presentation to the 16-bit data bus. Therefore, this part of the circuit converts the 8-bit data from the I/O device into 16-bit data for writing to the memory.

Glossary

Accumulator: A register that stores data to be used in an arithmetic or logical operation or the results of such operations.

Adder: A device that adds two binary numbers.

Address bus: A number of parallel lines that carry a binary pattern of 1s and 0s, which designate a specific memory or input-output location.

Adjacency: Two cells next to one another in the vertical or horizontal plane on a Karnaugh map are defined to be adjacent and may be combined.

ALU: An arithmetic-logic unit; a logic device capable of performing both arithmetic (add, subtract, and so on) and logical (AND, OR, and so on) operations.

ASCII: American Standard Code for Information Interchange.

Asynchronous circuit: A sequential circuit in which the next event occurs as soon as the current event is completed.

Base: The number of single digits in a number system.

BCD: Binary-coded decimal number system.

Binary: A number system using a base of 2.

Bit: An abbreviation for a binary digit.

Boolean algebra: A mathermatical algebra used to define logic operations.

Byte: A sequence of binary bits that is a submultiple of a binary word and is usually 8 bits.

Cell: A location on a K-map or the smallest element of a memory, capable of storing a 1 or a 0.

CMOS: Complementary Metal-Oxide-Semiconductor logic, logic constructed from both P and N MOS devices.

Combinational logic: A complex circuit formed by the interconnection of several gate circuits.

Comparator: A device that accepts two binary inputs and outputs information indicating their relative magnitudes.

Complement: To invert a binary word.

Core memory: A magnetic type of memory made up of miniature toroids.

Data bus: A number of parallel lines that transfer data to and from memory or I/O devices.

D-type flip-flop: A flip-flop whose data applied to its D input is clocked through to its output on reception of a clock pulse.

Decoder: A device for translating a combination of input signals into one signal that represents the combination.

DeMorgan's Theorem: The Boolean formulas $\overline{A} \cdot \overline{B} = \overline{A + B}$ and $\overline{A \cdot B} = \overline{A} + \overline{B}$.

Demultiplexer: A logic circuit that gates its input onto one of several outputs, depending on the binary word presented to its select inputs.

DIP: Dual-in-line package.

Don't care: A minterm for which the combinational output may be either a logic 1 or a logic 0.

DTL: Diode-transistor logic.

Dynamic Memory: Memory that requires refreshing in order to avoid losing its data.

ECL: Emitter-coupled logic.

Excess-3 code: A number system in which each word is equal to its binary equivalent plus 0011.

Execute: To perform an operation specified by an instruction.

Fan in: The number of inputs to a logic circuit.

Fan out: The number of inputs one gate's output is capable of driving.

Fetch: The phase of the instruction cycle in which the instruction is fetched from memory.

Flip-flop: A bistable multivibrator.

Full adder: A device that accepts two binary inputs and a carry in, adds them together, and provides a sum and carry output.

Gate: A circuit having two or more inputs and one output, the output of which depends on the combination of logic signals at the inputs. The output may be an AND, OR, NAND, NOR, NOT, EXCLUSIVE OR, or EXCLUSIVE NOR of the inputs.

Gray Code: A positional binary-number system in which each successive number differs by only one bit.

Half adder: A logic device that accepts two binary inputs, adds them together, and provides a sum and carry output.

Hardware: The electronic components of a digital computer.

Hexadecimal: A number system having a base of 16.

Instruction: A binary word that causes a computer to perform one or more of its operations.

Instruction register: The register in a processor that temporarily stores the instruction currently being executed.

Inverter: A device that accepts a logic 1 and outputs a 0, and accepts a 0 and outputs a 1.

JK flip-flop: A flip-flop with two conditioning inputs, J and K, and a clock input.

Karnaugh map: A display of a truth table in such a way as to allow the reduction of a Boolean expression.

Latch: A simple logic-storage element of the RS flip-flop type.

Least-significant Bit: The bit of a binary word with the smallest column weighting.

Logic: The science of dealing with the principles and application of truth tables, gating and so on.

LSI: Large-scale integration.

Master/slave flip-flop: A flip-flop composed of two internal flip-flops, one to receive the inputs (master) and one to drive the outputs (slave).

Maxterm: An OR of all the variables that together make up a function.

Memory: A device that stores binary data in locations called addresses.

Memory address register: A register in a processor that contains the address of the location currently being accessed.

Microprocessor: An integrated circuit that is capable of executing instructions.

Minimization: The process of reducing a Boolean expression so that the minimum number of gates is required to implement the function.

Minterm: An AND of all the variables that together make a function.

Modulus: The number of states through which a counter sequences during each cycle.

Monostable multivibrator: A circuit having only one stable state, from which it may be triggered to change state for a predetermined interval.

Most-significant bit: The bit of a binary word with the largest column weighting.

MSI: Medium-scale integration.

Multiplexer: A logic device that receives data on an input, selected by a binary code applied to its select inputs, and presents it to the output.

Negative logic: A form of logic in which the more-positive voltage level represents a logic 0 and the more-negative voltage a logic 1.

Nibble: A portion of a byte or a word, generally 4 bits.

Noise immunity: A measure of the insensitivity of a logic circuit to spurious or undesired signals.

Noise margin: The amount of noise a logic circuit can tolerate before giving a false output.

Octal: A number system using a base of 8.

One-shot: A monostable multivibrator.

One's complement: A means of representing a negative number by inverting the positive binary number.

Parity-bit: A bit that is set to a 0 or a 1 to make the total number of 1s in a binary word odd or even.

Positive logic: A form of logic in which the more-positive voltage level represents a logic 1 and the more-negative voltage a logic 0.

Product of sums: A Boolean expression consisting of a group of product terms ORed together.

Program counter: A register that contains the address of the next instruction to be fetched from memory.

PROM: Programmable read-only memory.

Propagation time: The time required for a level change at the input of a device to cause a change in level at the output.

Quine-McCluskey method: A method of minimizing Boolean expressions using tables instead of maps.

Radix: The base of a number system.

RAM: Random-access memory.

Refresh: The action of restoring data stored in memory to its original value.

Register: A device constructed from flip-flops that is capable of storing binary data.

Ripple counter: A counter constructed from flip-flops where the output of one stage is used to clock the next.

ROM: Read-only memory.

RS bistable: A flip-flop with a reset and a set input.

RTL: Resistor-transistor logic.

Schmitt trigger: A device with electronic hysteresis that is used to convert a slowly changing input waveform to an output waveform with sharp transitions.

Serial-to-parallel converter: A circuit such as a shift register that receives serial data and provides parallel outputs.

Software: Programs, codes, and other written information used with digital equipment, as opposed to the electronic equipment itself, which is referred to as the *hardware.*

Sum-of-products: A Boolean expression consisting of a group of sum terms ANDed together.

Synchronous logic: A logic system whereby all the elements are synchronized to a common clock.

Toggle: To cause a flip-flop to change to its opposite state.

Tristate logic: A gate that has three possible outputs: 0, 1, or "high impedance."

Truth table: A table that shows the output of a logic circuit for all possible combinations of input.

TTL: Transistor-transistor logic.

Two's complement: The one's complement of a number, plus 1.

Venn diagram: A diagram that represents AND and OR functions by intersecting circles.

Word: A binary number.

Bibliography

Alley, C.L. and Attwood, K.W. *Electronic Engineering.* New York: Wiley, 1966.

Gothmann, William H. *Digital Electronics.* Englewood Cliffs, N.J.: Prentice-Hall, 1977.

Hnatek, Eugene R. *A User's Handbook of Integrated Circuits.* New York: Wiley, 1973.

Dwyer, John. "Integrated Circuit Memories." *Wireless World,* April 1978.

Intel Corporation, *MCS 85 User's Manual.*

Markham, D.C. "Magnetic Bubble Memories, Part 1—The Device." *Electronic Engineering,* June 1979.

Mims, Forrest M. *Understanding Digital Computers.* Forth Worth, Texas: Radio Shack, 1978.

Raytheon Semiconductor Division, Digital. *Total Low-Power Schottky.* Mountain View, California.

Texas Instruments. *Characteristics and Applications of the SN7413N Dual Schmitt Trigger,* 1969.

Texas Instruments. *TTL: From the Beginning,* 1972.

Texas Instruments. *TTL Data Book for Design Engineers,* 1976.

Texas Instruments. *Schottky TTL.*

Walter, D.J. *Integrated Circuit Systems.* ILIFFE, 1971.

Watkin, R.V. *Computer Technology,* Longman, 1976.

Answers to Selected Problems

2.1 The output Q is high during time-periods E, G, and H, and low at all other times.

2.2 The output Q is high during time-period G only.

2.3 a. NOR
b. NAND
c. AND
d. EXCLUSIVE OR

3.1 1. A
2. $\overline{C}$
3. (A + B)C
4. $\overline{B}$

3.2 1. $\overline{B}C\overline{D} + BCD + A\overline{C}D$
2. $AD + ABC\overline{}$
3. $B(C + D)$

3.3 1. $f = \overline{\overline{AC} \cdot \overline{B}C}$, which may be realized using one 7400 package.
2. $f = \overline{\overline{A}\,\overline{C}} \cdot \overline{\overline{B}}$, which may be realized using two 7400 packages.

4.1 1. a. $1{,}271\frac{23}{32}$
b. $5{,}145\frac{65}{512}$
c. $3{,}345\frac{3}{8}$

2. a. $(2503.214)_8$
b. $(4516.1)_8$

4.1 a. 10111_2
b. 1100100_2
c. 10010001_2
d. 10110111.101_2
e. 10000111011.101_2
f. 101111100111.1101_2

4.3 1. $15\frac{7}{8}$
2. $16\frac{1}{8}$
3. $173\frac{5}{8}$

4.4 1. a. 010110001000_2
b. 111100010_2
c. 110000000011_2
2. a. 356_8
b. 363_8
c. 271_8
4.5 1. 111101_2
2. 1000010.000_2
3. 11101_2
4. 1010.10_2
5. a. 101000101_2
b. 1010000100_2
6. a. 101000_2
b. 11010_2
4.6 1. a. 001110000101_{bcd}
b. 100100100110_{bcd}
2. a. $D5A3_{16}$
b. $3CF6_{16}$
c. $35D7_{16}$
3. a. $E4_{16}$
b. 12567_{16}
4. a. $1F_{16}$
b. $74DA_{16}$
4.7 1. 1010_{XS3}
2. 1101_{XS3}
4.8 1. a. 1110_{TGC}
b. 1110110_{TGC}
2. a. 011001011_2
b. 0101101001_2

5.1 1.

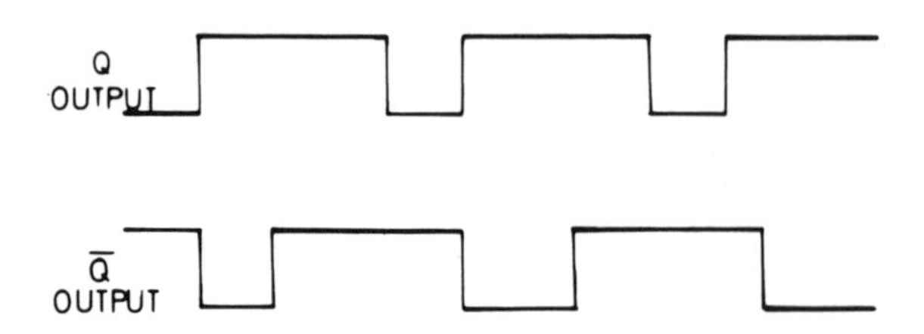

2.

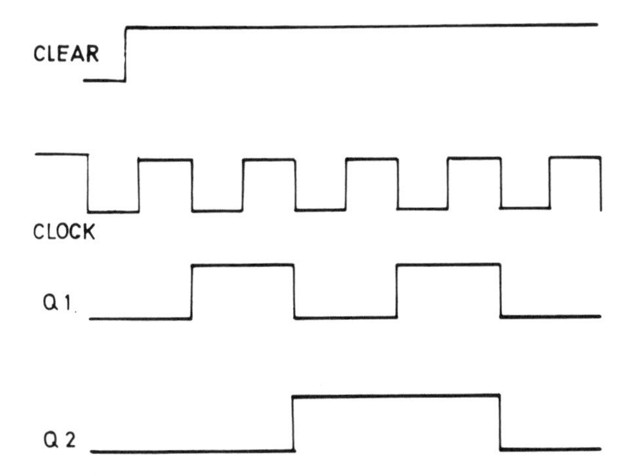

The circuit function is that of a divide-by-four.

3.

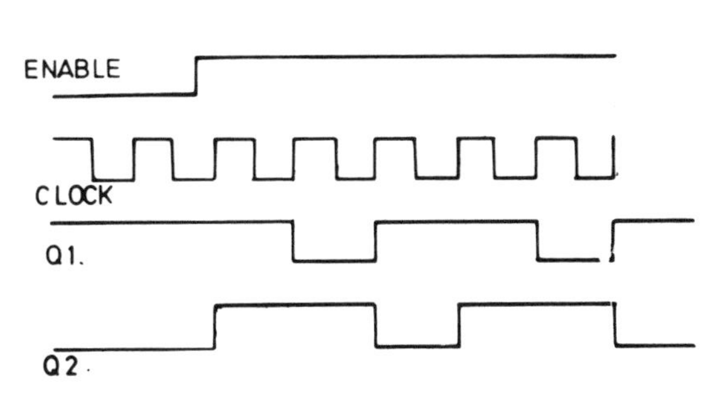

4.

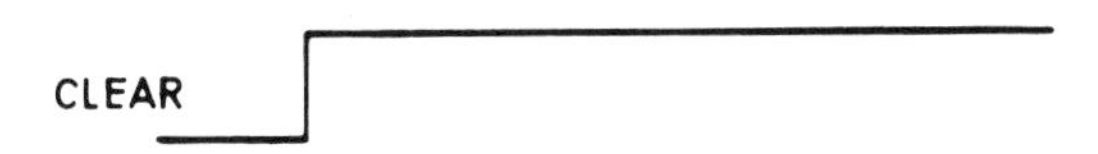

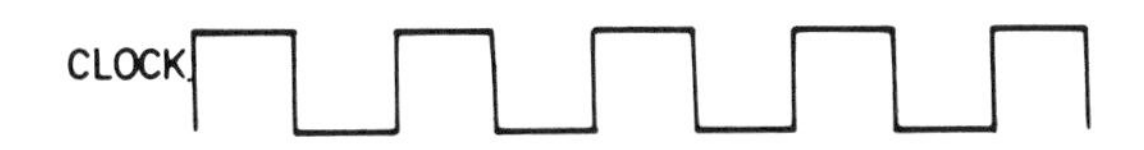

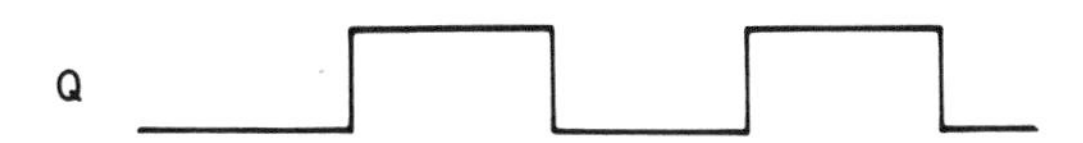

5.

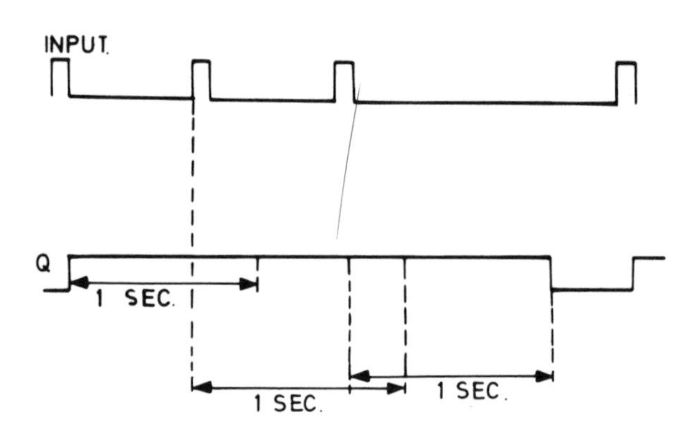

6.1 1. A modulo-13 counter may be obtained by constructing a modulo-16 counter and connecting a NAND gate to the Q_D, Q_C outputs (where Q_D = MSB and Q_A = LSB). The output from the NAND gate is then presented to the clear inputs of all four stages.

2. a. Construct a modulo-16 counter and connect a NAND gate to the Q_D and Q_C outputs (where Q_D = MSB and Q_A = LSB). Use the output of the gate to drive the clear inputs on the bistables.

   b. EXCLUSIVE OR adjacent outputs and EXCLUSIVE OR the most-significant output with a 0.

3.

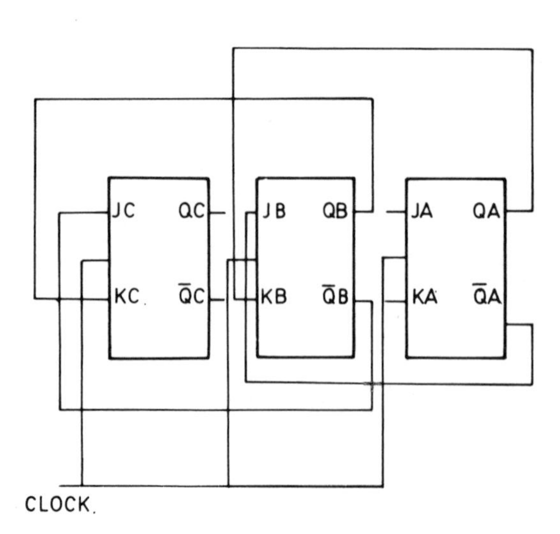

Index

Index

About the Author

John R. Scott is a customer training officer with Cossor Electronics Limited, Harlow, England. He received an honors degree in electrical and electronic engineering from the University of Nottingham in 1974. In 1975 he received the postgraduate certificate of education from the University of Cambridge and was then employed by Cossor as a service engineer before accepting his current position.